Feeding and Management of Livestock during Drought and Scarcity

Editors

Dr. N.V. Patil - graduated in 1983 from Nagpur Veterinary College, Nagpur and later obtained *M.V.Sc. (Animal Nutrition)* degree in 1985 from Post Graduate Institute, PKV, Akola. He obtained *Ph.D. (Animal Nutrition)* from NDRI, Karnal in 1993 and received Dr. S.K. Ranjhan best doctoral thesis award in 1993-95. He has 23 years experience of Teaching, Research and Research Management in various capacities at College of Veterinary Sciences and Animal Husbandry, Sardar Krushinagar, Livestock Research Station, Navsari in Gujarat Agricultural University, Central Institute for Research on Buffaloes, Regional Station, Nabha, Patiala and presently he is Head, Division of Animal Sciences and Forage Production and Director (acting) at *CAZRI, Jodhpur*. He has contributed, there are more than 198 publications.

Dr. B.K. Mathur - *B.V.Sc. & A.H (1980), M.V.Sc., Ph.D. (Animal Nutrition)* Principal Scientist, *CAZRI, Jodhpur,* started his career as Instructor, Veterinary College, Bikaner in 1983, and later served as Asst. Professor, Veterinary College, Gujarat Agricultural University and had practical experience of working in cattle feed industry before joining Agricultural Research Services in 1986. His major field of work is on utilization of locally available newer feed resources in ruminant ration mineral status of arid livestock and dissemination of feeding and health care technologies to the farmers, for the last twenty six years. He has worked for international project of ILRI, Ethopia, SDC, Switzerland and ACIAR-CSIRO, Australia, and is widely travelled. He has contributed more than 200 publications.

Dr. A.K. Patel- Principle Scientist, CAZRI, did *M.Sc. Dairying* (*Livestock Production and Management*) and *Ph.D. Dairying* (*Animal Genetics and Breeding*) joined Agricultural Research Services in 1990 as scientist at CAZRI, Jodhpur. His important contributions are in the field of improvement and conservation of livestock. His expertise is in designing the animal shelters for arid livestock and devising economically viable management system for optimum production in goats. He has over 100 pubications and working for national and international projects.

Dr. M. Patidar - Senior Scientist *(Agronomy)* and joined ARS in 1991 at CAZRI, Jodhpur. He has developed the silvipasture models for the arid region suitable for rejuvenation and improvement of dryland pastures and CPRs. He has published 25 research papers in different journals, 20 popular articles, 15 book chapters and 30 symposium abstracts.

Dr. A.C. Mathur- *B.Sc., B.V.Sc & A.H.* (1977), M.V.Sc. (Veterinary Gynaecology & Reproduction) from Indian Veterinary Research Institute, Izatnagar, Bareilly He worked as Assistant Project Officer at Semen Bank (1977 to 1980). He shifted to NDRI, Karnal to work in the leading RIA & Endocrinology laboratories with specified assay work at Bhabha Atomic Research Centre, Trombay, Mumbai. Joined CAZRI at its Krishi Vigyan Kendra, Jodhpur in the year 1984 as Training Associate (AH) is managing a modern Demonstration cum Research Unit of Tharparker breed cattle. He is associated with different research projects on livestock. He has published more than 90 articles.

Feeding and Management of Livestock during Drought and Scarcity

Editors

N.V. Patil
B.K. Mathur
A.K. Patel
M. Patidar
A.C. Mathur

Central Arid Zone Research Institute,
Jodhpur - 342 003

SCIENTIFIC
PUBLISHERS (INDIA)
www.scientificpub.com

Correct citation:

Feeding and Management of Livestock during Drought and Scarcity (*Eds. N.V. Patil, B.K. Mathur, A.K. Patel, M. Patidar and A.C. Mathur*) Pub. Scientific Publishers (India), Jodhpur

Published by:
Tanay Sharma
Scientific Publishers (India)
5-A, New Pali Road, P.O. Box 91,
Jodhpur – 342 001 (India)

E-mail: info@scientificpub.com
www.scientificpub.com

Print: 2012
Print, 2023

ISBN: 978-81-7233-647-9

Lasertype set : Rajesh Ojha

Binding at: Rajasthan Law Book Binding Works, Jodhpur
Printed in India

Foreword

Drought has been a frequent phenomenon both in the arid and semiarid regions and livestock has provided the much needed insurance to farmers during these periods of crop failure. However, the last fifty years have witnessed a tremendous rise in both human and livestock population. As a consequence, the nutritional support in the form of fodder and grazing resources has become scarce. This has been further aggravated because the climate change has increased the uncertainty and erraticness of rainfall as well as the intensity and frequency of droughts. These changes can alter both the living pattern and economic condition of dry region dwellers and, therefore, there is need for emphasizing the nutritional management of livestock.

Over the last three decades, livestock sector has grown faster than the crop sector as a whole, which has focused the attention of researchers and planners towards this sector for poverty reduction programs in most developing countries. In India, the growing livestock sector augurs well for the low income households as it leads to augmentation in their income and poverty alleviation. With the changing economic environment, livestock management will assume greater role to meet the growing demand of animal food products.

Although there is enormous possibility of fulfilling the targets of livestock related products in a demand driven economy, improving livestock productivity, its efficiency and quality remains a formidable challenge. Livestock production especially from dry areas, faces very atypical situations of a harsh environment, feed insufficiency both in terms of quantity and quality, and disease prevalence for which fortunately new technological options are now available but also need policy support. The extent to which the small scale farmers of the dry region will be benefited from livestock sector growth would depend on how the policies, technologies and institutions respond to their needs.

It is hoped that this book will provide a basis for the researchers and planners to understand the bottlenecks of animal production in drier areas and help them devise appropriate policies and programs for more pro-poor livestock sector growth. The contributions in this book have identified various researchable and policy issues for the benefit of farmers and the author need appreciation for highlighting them appropriately with possible technological options.

I congratulate all the authors and editors for bringing out this publication of addressing the issue of livelihood support to benefit livestock owners of the dry areas.

(A K Singh)

Deputy Director General (NRM),
Indian Council of Agricultural Research,
KAB-II, Pusa, New Delhi – 110 012

Preface

Livestock sector in India presently provides more than 25% of agricultural sector output, whereas it was 16% in 1970-71. The annual rate of increase has been impressive up to 4.3%, higher than the growth in agricultural sector of 2.8%. The livestock assumes greater dependable role in arid and semiarid areas which are frequented by the droughts of moderate to severe nature. However, animals' potential productivity is hampered by perpetuating problems of low quantitative and qualitative nutrition, lack of management efforts, breed degradation and low net returns due to unprofessional approach.

The livestock of this region on which millions of Indian farmers, landless laborers, market agents and their families depend for their livelihoods offer a major challenge and excellent opportunity for achieving economic growth. This can be achieved with the technological advancements and policy support in terms of breed up-gradation, nutritional security, economical management measures and proper health care supports.

The droughts which regularly visit this part of country, need to be addressed with preparedness and sound scientific technologies. The editors of the book were thus specifically requested to contribute towards sharing their experiences in resolving the problems of drought affected livestock and which will also have relevance towards the dry areas as a whole. The chapters of the book have been arranged in four major themes of livestock nutrition technologies, management efforts to address the harsh climatic factors, rejuvenation of scarce, limited resource base and management of health factors for improving productivity. The contribution requested from the experts included the scientists associated with livestock research in institutes located and working for dry areas like Central Arid Zone Research Institute (CAZRI), Central Sheep and Wool Research Institute (CSWRI), Rajasthan Agricultural University (RAU), Central Institute for Research on Buffaloes (CIRB) and Central Sheep Breeding Farm (CSBF). We look forward to get these inputs transformed into devising appropriate policies and programs for more pro-poor livestock sector growth and support the livelihood of the dry region livestock owners.

The editors are grateful to all authors for acceding to the request for contribution and hope that the book will help the researchers, technocrats and planners to harness potential of the livestock in the region.

N.V. Patil
B.K. Mathur
A.K. Patel
M. Patidar
A.C. Mathur

Contributors

A.C. Mathur Subject Matter Specialist, (A.H.), Krishi Vigyan Kendra, Central Arid Zone Research Institute, Jodhpur-342 003, Rajasthan

A.K. Gahlot Chairman-Faculty of Veterinary and Animal Science & Dean-College of Veterinary and Animal Science, Swami Keshwanand Rajasthan Agriculture University, Bikaner-344 001, Rajasthan

A.K. Patel Principal Scientist (Livestock Production and Management), Division of Animal Sciences and Forage Production, Central Arid Zone Research Institute, Jodhpur-342 003, Rajasthan

A.K. Shinde Principal Scientist (Livestock Production and Management), Division of Animal Nutrition, Central Sheep and Wool Research Institute, Avikanagar -304 501, Rajasthan

B.K. Mathur Principal Scientist (Animal Nutrition), Division of Animal Sciences and Forage Production, Central Arid Zone Research Institute, Jodhpur-342 003, Rajasthan

B.R.J. Mathur Ex.-Deputy Director, Animal Husbandry Department, Govt. of Rajasthan, Jodhpur – 342 001, Rajasthan

B.S. Punia Ex.-Director, Central Institute for Research on Buffalo, Hisar-125 001, Haryana

B.S. Rajpurohit Director, Central Sheep Breeding Farm, Hisar-125 001, Haryana

G.R. Purohit Ex.-Dean, College of Veterinary and Animal Science, Swami Keshwanand Rajasthan Agriculture University, Bikaner-344 001, Rajasthan

H.C. Bohra Principal Scientist (Animal Nutrition), Division of Animal Sciences and Forage Production, Central Arid Zone Research Institute, Jodhpur-342 003, Rajasthan

Harpal Singh Head and Principal Scientist (Agricultural Structure and Process Engineering), Division of Agricultural Engineering and Energy, Central Arid Zone Research Institute, Jodhpur-342 003, Rajasthan.

J.P. Singh Principal Scientist (Economic Botany), Regional Research Station, Central Arid Zone Research Institute, Bikaner-344 004, Rajasthan

J.R. Sharma Project Director, Regional Remote Sensing Service Centre, ISRO, Department of Space, CAZRI Campus, Jodhpur-342 003

L.N. Harsh Principal Scientist (Forestry), Division of Integrated Land Use Management and Farming System, Central Arid Zone Research Institute, Jodhpur-342 003, Rajasthan

M. Patidar Senior Scientist (Agronomy), Division of Animal Sciences and Forage Production, Central Arid Zone Research Institute, Jodhpur-342 003, Rajasthan

M.K. Mandape Project Director, ZCU-VI (ICAR-TOT), CAZRI Campus, Jodhpur-342 003, Rajasthan

M.P. Rajora Senior Scientist (Plant Breeding), Division of Plant Sciences and Biotechnology, Central Arid Zone Research Institute, Jodhpur-342 003, Rajasthan

M.S. Khan Principal Scientist (Biochemistry), Division of Animal Sciences and Forage Production, Central Arid Zone Research Institute, Jodhpur-342 003, Rajasthan

Manjit Singh Director, National Research Centre for Mushroom, Solan-173213, Himachal Pradesh

N.V. Patil Director (acting), Head and Principal Scientist (Animal Nutrition), Division of Animal Sciences and Forage Production, Central Arid Zone Research Institute, Jodhpur-342 003, Rajasthan

P.C. Lailer Senior Scientist(Livestock Production and Management), Central Sheep Breeding Farm, Hisar-125 001, Haryana

P.P. Rohilla Head (acting) and Sr. Scientist (Livestock Production and Management), Regional Research Station, Central Arid Zone Research Institute, Pali-Marwar- 306 401, Rajasthan

Pratap Narain Vice Chancellor, Swami Keshwanand Rajasthan Agriculture University, Bikaner-344 001, Rajasthan

R.K. Beniwal Head and Principal Scientist (Soil Science), Regional Research Station, Central Arid Zone Research Institute, Bikaner-344 004, Rajasthan

S. Kachhawaha Subject Matter Specialist (Veterinary Science), Krishi Vigyan Kendra, Regional Research Station, Central Arid Zone Research Institute, Pali-Marwar-306 401, Rajasthan

S.A. Karim Director, Central Sheep and Wool Research Institute, Avikanagar -304 501, Rajasthan

S.K. Kaushish Ex.-Principal Scientist (Livestock Production and Management), Division of Animal Sciences and Forage Production, Central Arid Zone Research Institute, Jodhpur-342 003, Rajasthan

S.S. Paul Senior Scientist (Animal Nutrition), Central Institute for Research on Buffaloes, Sub-Campus, Bir Dosanjh, Nabha-147 201, Patiala, Punjab

T. Sharma Head (acting) and Associate Professor (Animal Nutrition), College of Veterinary and Animal Science, Swami Keshwanand Rajasthan Agriculture University, Bikaner-344 001, Rajasthan

T.K. Bhati Principal Scientist (Agronomy), Division of Integrated Land Use Management and Farming System, Central Arid Zone Research Institute, Jodhpur-342 003, Rajasthan

V.S. Rathore Senior Scientist (Agronomy), Regional Research Station, Central Arid Zone Research Institute, Bikaner-344 004, Rajasthan

V.S. Shekhawat Divisional Manager, New India Assurance Company, Udaipur, 313 001. Rajasthan.

Contents

PRODUCTION MANAGEMENT

HEALTH MANAGEMENT

1

Feeding practices of livestock during drought conditions-importance of feed supplements

B.K. Mathur

Central Arid Zone Research Institute, Jodhpur-342 003

Livestock plays a crucial role, contributing through milk, meat, wool and their products. Materialization of optimum utilization of feed and fodder resources to combat scarcity, needs practical strategies through planners to enhance livestock productivity in India, and other developing countries. Livestock productivity is becoming a major consideration for many developmental programmes. According to FAO estimate of 2002, the country had 219.6 million cattle, 80.6 million buffaloes, 123.5 million goats, 58.2 million sheep, 17.5 million swine and 413 million poultry. Population figures indicate that Indian cattle, buffaloes, sheep and goats constitute 16, 54, 5 and 21% respectively of world population. Put together these livestock contribute 1830 billion rupees to GNP, which in turn is 32% of 27% contributed by the agriculture.

In arid zone traditional rural wisdom prefers livestock based agricultural practices as compared to crop farming. Most of the research findings with holistic perception in the field of agriculture and animal husbandry also confirm it. The recent survey carried out by NIANP, Banglore indicate that there is shortage of 45% of dry roughages, 44% of concentrates and 38% of green fodder (Anon. 2001-02). Arid zone is facing continuous drought since last four years. However, the recent drought (2002-03) is the severe most due to cumulative effects of preceding years. When crop fails, livestock serves as a ready source of revenue for the livelihood of arid farmers. Thus livestock is the backbone of rural economy in the Thar Desert.

As per the latest livestock census (2003), Rajasthan state harbours 49.1 million livestock, out of which arid zone (12 districts) livestock population is 28.60 million. The native livestock breeds of arid zone have inherited through generations' adaptability to adverse climatic conditions of the hot desert.

During drought the first and foremost branch which needs utmost attention is animal nutrition or the feeding management.

FEEDING MANAGEMENT

Due to the acute scarcity of feed and fodder during drought the utmost importance is given to the feeding component. Dried roughages provides bulk to the belly and meets 70-80% supply of dry matter and total digestible nutrients (TDN) required by cattle, sheep and goats. Failure of the monsoon is the main reason responsible for drought in the country. Arid areas where mostly rain fed agriculture is practiced are most prone to fodder scarcity than others.

Traditionally, the arid zone farmer conserves and stores the bajra straw, grasses etc., sometimes blended with legume crop byproducts of mung and moth in a typical dom shape storage, called "Bhanwariya" in local language. The walls are constructed using less edible crop byproducts and vegetation. These fodder stores are opened during drought to maintain the livestock for few months. During present continuous drought situation livestock keepers have already exhausted stocks long back and have nothing to feed to the livestock.

Status of feed and fodder resources: There is an evident scarcity of feed and fodder resources in arid zone. In a case study, it is observed that in villages of Shergarh tehsil on an average had a deficit of fodder ranging from 21 to 28% (Mathur et al., 2003) Where as in Solankiyatala and Sai villages of Shergarh tehsil the deficit was to the tune of 80% it clearly shows that due to cumulative effect of drought from year 1998 to 2000, and 2002, there is sharp decreased in livestock population. Livestock decreased by 65 and 62% in the above villages due to lack of feed, fodder, sale of animal, high morbidity and mortality rate. In both the villages the livestock owner has no store of feed and fodder due to continous drought. Fodder is being transported from neighboring state and made available to them on cost by local bodies, but the quantity and frequency is irregular. In the common grazing land (CGL) the edible grasses, shrubs/ bushes has almost disappeared (Mathur et al., 2002). Some of the factors responsible for this deficiency, such as degradation of pasture, drastic reduction of CPRs and persistent drought resulting often into complete crop failure and consequently complete crop residue failure. Most of the feed and fodder ingredients are now being imported from other areas and states. Some of the dietary ingredients such as ground nut roughage and wheat straw have replaced the erstwhile *Lasiurus sindicus* straw and *Zizyphus numularia* leaves almost to the extent of 80%. Similarly, the traditional ingredients of concentrate ration, such as, sesame cake and Guar (*Cymopsis tetragonoloba*) seeds have almost been completely replaced by cotton seed/ground nut/mustard cake, legume churi (coarse ground industrial by product), DORB, wheat bran etc. This switch over has resulted into higher incidence of metabolic (production) diseases. Production diseases were earlier considered to be related to only high producing exotic/cross bred animals. Their occurrence in the low producing indigenous animals has proved that dietary factors have become more important than the high production. Grazing livestock are commonly infested with various endo parasites leading to clinical or sub clinical manifestation and enormous lost to livestock owner both direct as well as indirect as. Digestibility of nutrients also reduces lowering feed utilization. Due to pathogenic changes the animal become weak with retardation in growth, production and working ability (Panda and Baidya, 2004).

Deficiency of nutrients: Several deficiencies have been identified in the animals of arid zone (Dongre, 2000) because of evidence of existence of deficiency. Energy, protein and mineral deficiency results in to pica. Deficience of calcium, phosphorous, zinc, manganesium, copper, cobalt, iron etc., are not only prevailing unabated but are also registering an upward trend, with increasing reproductive disorders, particularly anestrus in arid region (Mathur *et al.,* 2001).

Under such condition meeting of fodder demand by transportation of wheat straw or paddy straw etc., from neighboring states becomes a major task. Moreover, the roughages like wheat/paddy straw are of very poor nutritive value in terms of digestible crude protein (DCP) and total digestible nutrients (TDN). These straws as such are unable to meet even the maintenance requirement of livestock and thus lead to higher rates of nutrition related health problems. Similarly, the conventional concentrate feeds like oil cakes, legume and cereal grain or their byproducts are available at unaffordable high prices and becomes out of reach. During this period resources like unconventional feeds, non protein nitrogen and mineral mixture will play an important role when utilized judiciously.

Therefore, during drought following techniques for livestock feeding will be beneficial for the well being of animals.

(A) Preparation of ration with urea-molasses for immediate feeding (Per cattle/day)

1. Wheat/paddy straw 3.0 kg.
2. Molasses 300 gms.
3 Urea 30 gms.
4. Vitamin Mineral Mixture 20 gms.
5. Common salt 20 gms.

Procedure of Feeding

In a container dissolve 30 gms urea in half litre of water and add 300 gms molasses, stir it with a stick homogenously. Now spray evenly this urea molasses solution on straw and then also spray on straw vitamin mineral mixture with common salt. Finally using both hands mix the ingredients with straw thoroughly. Ration is ready to be fed to livestock. For making ration of 100 kg straw we require one kilogram urea and ten kilogram molasses. This ready to fed ration is very palatable to animal and is consumed immediately.

(B) Urea treatment of poor quality straw

Wheat and paddy straw is crop residues lift after harvesting the grains from crops. These straw forms the main bulk of roughages for cattle during drought. These cereal roughages are poor in nutritive value, with almost 0% digestible crude protein. Due to high lignin content the palatability is also low. High oxalic acid content in paddy straw affect calcium absorption. So when as such dry straw is fed to cattle the maintenance requirement is not met with. Therefore, with dry straw either come green fodder or grain/cake is mixed and offered to cattle. But under drought circumstances, these are not available with livestock keepers. Keeping this in view the alternative to be adopted is 3% urea treatment of the poor quality straw with 50% moisture level and stored under cover for 21 days. This urea treated straw can meet the maintenance requirement of the animal to a large extent.

Procedure of urea treatment of straw:

Material required: For 100 kg straw we require 3.0 kg urea, 40-50 litre water, buckets, polythene sheet or used empty polythene bags.

Method

- First spread 25 kg straw layer on clean pucca cemented floor in 3x5 feet area.
- Dissolve 3.0 kg urea in 40-50 litres of water.
- Now spray 8-10 litres of urea mixed water on spreaded straw and compress evenly with feet by 2-3 persons.

- Over the spreaded straw similarly put three more layers of about 25 kg straw one over another layer. While putting on layer of straw over another each time spray 10 litre of urea dissolved water and press it with feet to minimize air entrapped between straw.

- Now cover the treated straw with polythene sheet or bags in a air tight manner so that near anaerobic condition can be maintained.
- Put some weight over the covered treated straw like stone slats etc. to keep polythene intact.
- After 21 days the enriched treated straw is ready to feed the livestock.
- This treatment increases the nutritive nature of straw in terms of protein and energy many folds. Since the lingo-cellulose bond is broken, the protein and cellulose of the straw in addition to urea nitrogen will be made available to animal.

Precautions

(a) Urea mixed water should be used immediately and must be kept away from livestock because; as such it will be highly toxic to animals.

(b) Spray urea mixed water evenly on straw.

(c) Feeding of Urea treated straw should be increased slowly day by day in the ration of animals to develop its palatability.

Commonly used poor quality roughages in India

Crop Residues and agro industrial by products			
1.	Wheat straw	12.	Sorghum stover
2.	Rice straw	13.	Millet stover
3.	Bajra straw	14.	Cotton by products
4.	Masoor straw	15.	Cotton straw
5.	Sugar cane tops	16.	Fruit canning waste
6.	Corn stover	17.	Rice hulls
7.	Corn cobs	18.	Sugar cane bayasee
8.	Oat straw	19.	Cotton seed hulls
9.	Sunflower stover	20.	Coffee beanhull
10.	Cassava wastes	21.	Soyabean straw
11.	Cassava leaves	22.	Banana pseudo stem

It is an established fact that these poor quality roughages are less palatable and nutritionally poor than cultivated grasses or legumes commonly used in different animal production in their nutritional quality. Among the several roughage sources available cereal straws with an annual total availability of about 250 million tonnes (Mudgal ad Pradhan, 1989) received maximum attention for improving their palatability and nutritive value because of their huge availability. The significance of improved animal nutrition on productivity and its continuing constraint on prevailing animal production systems has been repeatedly focused across the developing countries (ILRI, 1995). Inspite of huge gap existing between availability and demand for dry matter energy and protein at national level for feeding different species of livestock, still in states like Punjab and Haryana cereal straws more specifically rice straw is being burnt in the fields contributing to loss of valuable energy source to the national pool. Under global context wheat straw (25.3%) is the most abundant crop residue followed by rice straw (22.8%), corn stover (15.8%), barley straw (8.3%), sugarcane tops (8.1%) and sugarcane by products (6.3%) as reported by Han and Garett (1986). Asia is producing about 46% of the world supply of crop residues; therefore Asian Animal Nutritionists have paid more attention to improve the nutritive value and utilization of crop residues as animal feed in relation to other regions. Ninety two percent of rice straw and thirty four percent of the wheat straw are produced in Asia.

Many methods of treatment have been exhaustively investigated before accepting urea ammoniation as the most potential method for field scale application in most of the Asian countries. Urea treatment can be done in different ways, depending on the local conditions and preferences but some rules can be given regarding concentration of urea, duration of treatment, amount of water tq be used and way of stacking.

Table: Chemical composition of lignocellulosic by product

Nutrient	Wheat straw	Rice straw	Sugarcane tops	Oat straw	Rice hulls	Corn stover
Cellulose	36.3	37.0	25.7	36.5	42.0	31.0
Hemicellulose	24.4	21.1	28.8	31.8	18.0	28.5
Holocellulose	60.7	58.1	54.5	68.3	60.0	59.5
Lignin	9.7	8.2	7.5	8.5	26.0	13.0
Crude protein	2.6	2.1	6.3	3.4	2.0	6.0
Ash	5.8	8.4	5.0	3.6	12.3	7.0

(C) Chopping of straw

For feeding livestock always offer chaffed fodder. Chopping improves digestibility and avoids nutrient wastage.

(D) Addition of leguminous crop by products

Feeding of wheat/rice straw by mixing with leguminous straw e.g., gram straw, groundnut crop byproduct or masoor straw will improve nutrient availability and digestion of the livestock.

SUPPLEMENTARY FEEDING

1. ***Use of urea molasses mineral blocks :*** Supplementation of urea molasses blocks as lick along with straw feeding is recommended during drought. Block contains urea, molasses, minerals along with some grain/cakes and a binder. Animals meet out their requirement of energy, protein and minerals by licking these blocks. The blocks

available in market are compact, normally weights around 3 kg. And occupies less space and are easy to transport.

2. ***Providing vitamin "A" doses*** : Provision of vitamin "A" sources in animal ration is essential. Green fodder contains carotene which is being converted into vitamin "A" in animal body. However, during drought green fodder is not available to livestock and thus animal shows symptoms of vitamin "A" deficiency which are predominantly as under:
    - Lachrymal discharge
    - Loss of resistance power to diseases.
    - Frequent diarrhea.
    - Corneal opacity.
    - Newly born calves, lambs are blind.

Treatment: -

Provide vitamin A injection at bimonthly interval and/or in ration add vitamin A containing mineral mixture.

Vit A 440 *I.U.*/kg B.wt

Parentral injection of aqueous solutions than oily solutions

Excess doses cause – rarefaction of bones and fetal abnormalities

Control : daily requirements

Vit A 30 *I.U*/kg *B.wt*

Carotene 70 *I.U.*/kg B.wt

3. Cheaper and locally available unconventional feed resources : Arid region has many unconventional feed resources in the form of agricultural produce, weeds, byproducts from agro-processing methods etc. some of these have been used in the feeding of livestock and found to be useful to reduce cost of feeding without any harmful effect on health and production function of livestock.

Tumba (*Citrullus colocynthis*) seed cake - Tumba plant, an annual creeper, grows naturally in abundance in hot arid areas of the country with minimum possible water availability. It grows and multiplies very fast during monsoon season and its fruits; Tumba fruits are available in the month of October-November. Mature fruits are golden yellow in colour, ball like, 10-12 cm in diameter.

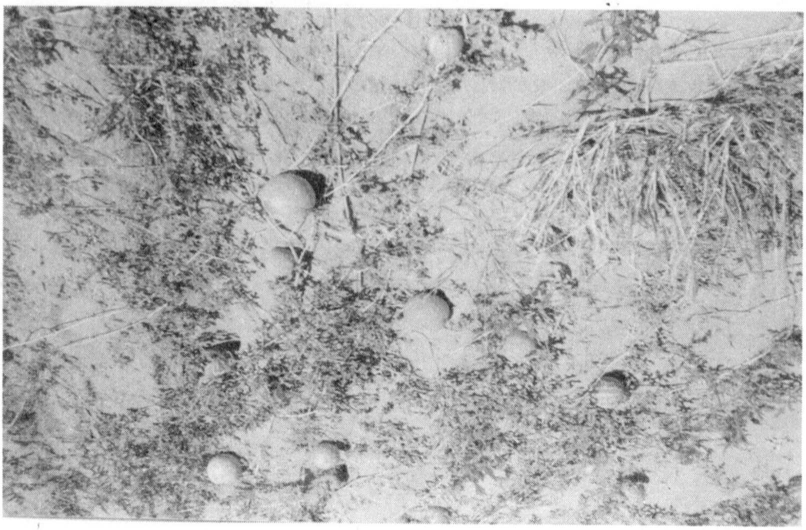

The taste of the fruit is very bitter. Tumba seed are rich in fat, having up to 16% oil. The Tumba (*Citrullus colocynthis*) seed cake (TSC) a by product of the oil extraction industry is nutritionally rich as it contains 16 to 22% CP.

Presently, TSC is available in abundance in arid regions and is being used as a fuel for furnances in factories, and its thus wasted.

A simple practice of 25% inclusion of tumba seed cake (TSC) in concentrate feed found lower the cost of animal feeding by 18% to 35% without having any adverse effect on the production, general health and reproductive performance.

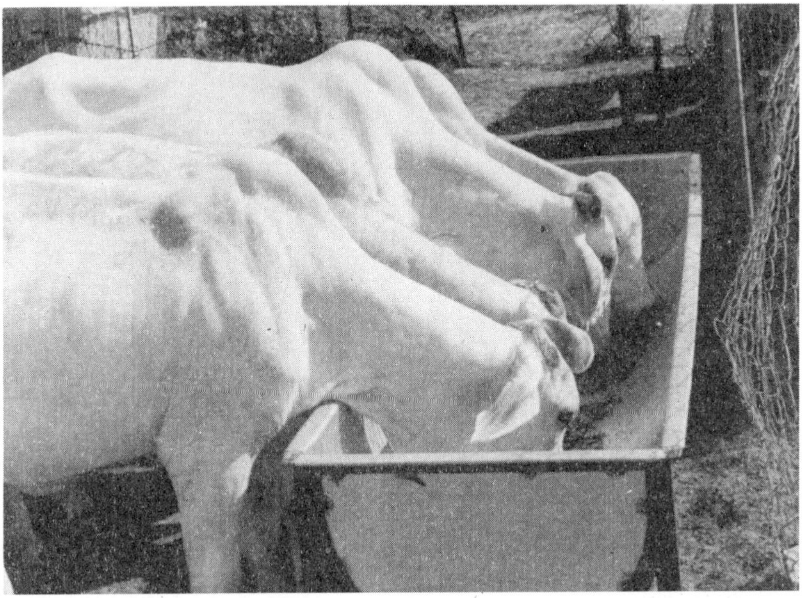

In addition, it would result in the utilization of locally available non-conventional cheaper protein source, abriding the gape between demand and supply of the scarce protein, thus, benefiting the marginal farmers appreciably (Mathur *et al.*, 2000, *Mathur et al., 1989, Mathur, 1996*).

Lani (*Salsola baryosma*): Salty shrub Lani (*Salsola baryosma*) of arid region available at vegetative stage can be a good source of fodder to animals. Shoots consists of mainly fleshy stem, since leaves are very minute, consisting not more than 5- 7% of the total biomass. The acceptability and palatability of fresh cut shoots showed that Lani is palatable (*Mathur et al., 2007*).

Lana (*Haloxylon salicornicum*) seeds: Lana (*Haloxylon salicornicum*) salty weed available in the arid region produces seeds having CP 18.60% found to be good source of protein in the concentrate part of lactating cows (Anonymous, 2007), goats and camel replacing about 25% of the conventional sesamum cake.

Lana (*Haloxylon salicornicum*) Arid feed resource for Camel, Goat and Cattle

***Colophospermum mopane* leaves:** Although sole feeding of fresh *C. mopane* leaves is not possible in goats but it can very well be fed as a part of complete rations at 35-40% of whole ration (Patil *et al.*, 2007). But it was observed that the palatability of dry *Colophospermum mopane* leaves was low and decreased with progress of feeding from 15 to 5% from 1st to 5th week and the traditional local P cineraria leaves were better source of supplementation even in the dry form supporting the milk yield of goats. (*Mathur et al.,* 2006).

***Hardwickia binata* as a Supplement in Goats:** *H. binata* is available in plenty in the arid region having rainfall above 300 mm. the study showed that the fresh leaves are palatable to goats and sole feeding supports the body weight growth of growing kids (Patil *et al.*, 2009).

Feeding Vilayati Babool (*Prosopis juliflora*) Pods Powder: PJPP can be used up to 35% in the concentrate of lactating goats with maintenance of milk production without having any detrimental effect on health (Mathur *et al.,* 2003).Similarly *P. juliflora* pod husk can be used up to 50 percent level in the concentrate along with tumba (*Citrullus colocynthis*) seed cake as low cost ration in the sheep without any adverse effect on animal health (Mathur *et al.*, 2002).

Thornless Cactus (*Opuntia ficus indica*) : A new fodder source *Opuntia ficus* indica- a thornless cactus introduced in Indian arid region was found to have the fodder value as a maintenance feed and was observed to reduce the water requirement if fed along with the dry roughages in goats, sheep and growing cattle. In addition its high mineral content may reduce the mineral requirement, as arid animals are suffering from mineral imbalances (*Mathur et al., 2009*).

Utilization of monsoonal weeds of arid zone : The monsoonal weeds available in the rocky and sandy habitat of arid region have a distinct relative preference index for grazing sheep and goats and it was studied to be in the order of 1. Kanti (*Tribulus terristris*) 2. Kagio (*Tetrapogon tenellus*) 3. Santo (*Trianthema protulacastrum*) 4. Lolaru (*Digeria muricata*) 5. Bekario (*Indigofera cordifolia*) and 6. Gangan (*Grewia tanax*) for varieties tested in the region. (*Patil et al., 2005*).

4. Feeding of Mineral mixtures and common salt : Feeding managers of animals should have mineral bricks and common salt, so that deficiency of minerals and common salt may not occur. Calcium, phosphorus, copper and mangnesium was significantly low in livestock of villages in arid zone of western Rajasthan. Pica in cattle, buffaloes, camels and goats in this area is another problem related to mineral deficiencies (*Mathur et al., 2005, Mathur et al., 2009*). Milk fever in this area occurs mainly due to calcium deficiency only within one week post-parturient having sufficient level of magnesium. The infertile cattle bears a history of under fed with malnutrition, and per rectum examination of genitalia revealed hypoplastic and/or smooth ovaries and persistent corpus luteum was very common, and they responds to mineral mixture – vitamin supplementation. There is need of area specific mineral mixture supplementation.

5. Deworming : Animal should be dewormed with broad spectrum anthelmentic eg., Albendazole/Fenbendazole/ Closentol etc. thrice yearly. In case of ectoparasites: spray/ dusting should be done at regular intervals with ectoparasiticide. These simple treatments will increase the availability of nutrients to livestock.

CASE STUDY

Survey was conducted in four villages during drought, namely, Sointra, Shergarh, Sai and Solankiyatala of Shergarh Tehsil, district Jodhpur of western part Rajasthan, India. Data were collected with a specially designed interview schedule, group interview, field animal parameters and through participatory rural appraisal. A cross section of the population living in scattered settlement called "Dhani" were interviewed.

It is observed that in village Sointra and Shergarh on an average had a deficit of fodder ranging from 21 to 28% and that of concentrate it is from 48 to 79 percent. While in Sai and Solankiyatala the fodder and concentrate deficit is upto 80%. The average grazing distance traversed (kg/day) in cattle varied from 1.8 to 4.8 km, sheep it was as high as 10.0 kms while goat traversed between 4 to 6 kms. Incidences of diseases were classified system wise viz gastro intestinal (GIT), respiratory (RIT), reproductive and skin. The occurrence of gastro intestinal diseases was highest (29%) in Sointra, reproductive (26%) in Shergarh, and skin (28%) in Sai followed by Solankiyatala (25%) was recorded.

Reproductive parameters studied among the 20 households surveyed in each village depicts that age of maturity in cattle were higher in Sointra and Shergarh villages as compared to Sai and Solankiyatala which was 48 and 41-35 months respectively. The reason understood may be that minimum gastrointestinal diseases were reported in these village reflecting comparatively good feeding. The other reason may be the percent population of cattle in surveyed households was very less (7 to 16%) of the total livestock population while sheep and goat population was more than 80% of the total livestock numbers. Anestrous cases were higher (15-25%) in Solankiyatala village. Morbidity rate was highest in Shergarh village due to GI disturbances 29% followed by Sai village due to skin 28%, then in Shergarh with reproductive disorders 26%, Solankiyatala with skin affection 25% and then of respiratory infection 22% in Sointra village. High GIT and RTI probably is responsible for higher age of maturity in cattle (Mathur et al., 2004).

Thus, it is concluded that in the Surveyed villages there is scarcity of feed and fodder. The nutrient requirement of the livestock needs to be meet up by balanced feeding and to build up the resistance of livestock against diseases. Water channels needs to be constructed in CPR with routine water supply connection. Planning is required to improve the common properly resources by resowing grasses, shrub and top feeds.

Feeding management of livestock during drought can be summarized as follow

- Preparation of ration with urea-molasses for immediate feeding
- Urea treatment of straw
- Chopping of straw
- Addition of leguminous crop by products

Supplementary feeding

- Providing Mineral vitamin mixture
- Use of urea molasses mineral blocks
- Providing vitamin "A" doses
- Cheaper and balanced concentrate for lactating animals
- Mineral bricks and common salt

Deworming with broad spectrum anthelmenthic.

Observations recorded during drought condition in arid villages

- Acute scarcity of feed and fodder
- stored fodder has been exhausted long back
- no feed fodder available in community grazing land
- daily grazing distance has increased significantly
- even the population of *caparis decidua* reduced considerably
- health of arid livestock adversely affected
    - weak
    - debilated
    - consuming inanimate objects
    - ruminal atony
    - pneumonia
    - skin infestation
    - abscess around neck

Reproductive disorders:

- increase in age of maturity

- abortion
- prolapse
- anoestrus

Sequelae: high morbidity and mortality rate

Possible measures

- Maintenance requirements of livestock need to be met : suggested and demonstrated poor quality fodder enrichment and supplementation with balanced concentrate.
- Veterinary assistance needs strengthening
- Water points need to be maintained
- Fodder transport and distribution needs improving from near by state: straw- mostly having 0.00% DCP and low TDN
- Common Grazing Lands need improvement
- People should be encouraged to grow fodder on their land

Livestock Parameters to be studied during drought condition

- Feed resources available
    - fodder deficit
- Grazing distance traversed
- Occurrence of diseases
- Reproductive parameters
    - Age of Maturity
    - Anestrous
- Morbidity disease

References

Anonymous (2007). Feeding Lana (*Haloxylon salicornicum*) and Saji (*Haloxylon recurvum*) to Lactating Rathi Cattle. Annual Report, CAZRI 2007, pp.101-102.

Anon 2001-2002 Annual report National Institute of Animal Nutrition and Physilogy Bangalore, Karnataka.

Banerjee, G.C. (1998). A text book of Animal Husbandry. Publishing by Raju Primalani for oxford & IBH Publishing Co. Pvt. Ltd. 66 Janpath New delhi 110 001. p, 1067.

Block, E. 1984. Manipulating dietary anions and cations for prepartum dairy cows to reduce incidence of milk fever. J. Dairy Sci.67:29-39.

CAZRI 2001. CAZRI Annual Report. Central Arid Zone Research Institute Jodhpur-342003.

Dhami, A. J., Patel, P.M., Lakum, P. D., Ramani, V. P. and Pande, M.B. 2003. Micro nutrient profile of blood plasma in relation to age reproductive status of Holstein Friesian cattle. *Indian J. Anim. Nutr.* 20 (2) 206-211.

Dongre, R.A. (2000).Haemato-biochemical studies in relation to trace mineral status in camel.M.V.Sc.thesis submitted to the Rajasthan Agriculture University, Bikaner (Rajasthan).

F.A.O. YEAR BOOK 2002. Rome, Italy.

Garg, M. R, Bhandari, B. M. and Sherasia, P. L. 2003. Macro and micro – mineral status of feeds and fodders in Kota district of Rajasthan. *Indian J. Anim. Nutr.* 20 (3) 252-261.

Gehlot, A.K. (2006). Animal health management in drought prone areas. Paper presented in seminar on" *Sustainable livestock production in arid region*". 22-23 May 2006 held at College of Veterinary and Animal Sciences, Rajasthan Agriculture University, Bikaner.

Han I K and Garrett W N 1986. Improving the dry matter digestibility and voluntary intake of low quality roughages by various treatments : A review p. 200.

ILRI 1995. Global agenda for livestock research (Eds. P. Gardinar and C. Devendra). Processing of a consultation, ILRI. Nairobi, Kenya. P.118.

Kaur, Harjit and Panda, N. 2004. Role of micronutrients in relation to immunity and reproduction. Proc. XI Animal Nutrition Conference, Jabalpur, 188-194.

Krishna N (2003). Processing of roughages for improved utilization. Paper presented in the workshop on "Current status of feed processing and future strategies" 22-23rd July 2003, held at ANGRA Agricultural University, Rajendranagar, Hyderabad, 500 030.

Krishna, N. and Prasad, D.A. 1991. Rumen degradability of crop residues in Small Ruminants. Proc. National Seminar held at Tirupati on 'Small Ruminants Production in India by the year 2000 – Strategies for Development' during 13-17 November, 1990. pp. 103-113.

Livestock census (2003). Animal Husbandry Department, Government of Rajasthan, Jaipur (Rajasthan).

Mathur, A.C., Mathur, B.K., Kaushish, S.K. and Patel, A.K. 2001. Effect of package of improved nutritional management practices on true anestrus cases in buffaloes in several arid Rajasthan. Paper presented at Annual Convention and National symposium on `Fertility Managment of Farm animals under advance agroclimatic conditions, Oct. 6-8, 2001, Jodhpur pp. 39.

Mathur B.K. 1997. Tumba seed cake invented as cheaper feed for sheep. Carpet-e-world. XIX 88-89.

Mathur B.K. and Mittal, J.P. 1987. Deficiency diseases of livestock in village conditions of western Rajasthan. Scientific programme and abstracts [1(a)11]. National symposium on Livestock production vis-à-vis fodder and pasture development in western Rajasthan, 2-3 March 1987, Livestock Research Station, Chandan (Jaisalmer-Raj).

Mathur, B.K., Siyak, S.R. and Paharia, Shailesh (2002). In corporation of *Prosipis juliflora* Pod husk as Ingredient of low cost ration for sheep in arid region. Current Agriculture 26:1-2, 95-98.

Mathur, B.K., Siyak S.R. and Bohra, H.C. (2003). Feeding of Vilayati babool (*Prosopis juliflora*) pods powder to Marwari Goats in Arid Region. *Current Agriculture. 27(1-2):57-59.*

Mathur BK, 1996. Effect of replacement of cotton (*Gossypium* spp.) seed cake from Tumba (*Citrullus colocynthus*) seed cake and sun dried poultry droppings on the performance of sheep. Ph.D. thesis submitted to Rajasthan Agriculture University, Bikaner, Rajasthan.

Mathur, B. K. (2004). Drought proofing through utilization of low – grade roughage and non-conventional feed resources for small ruminants. National Seminar on Opportunities and Challenges in Nutrition and Feeding Management of Sheep, Goat and Rabbit for sustainable production. (February, 10-12). Organized by *Indian Society of Sheep and Goat Production and Central Sheep and Wool Research Institute* Avikanagar-304501, Rajasthan. pp. 191-201.

Mathur, B.K., Ahuja, Usha Rani., Kumar, Suresh, Saha, D.K., Narain, P., Friedal, M.F. and Bastin Gary. (2004). Available feed resources and sheep and goat health status on degraded arid rangelands. National Seminar on Opportunities and Challenges in Nutrition and Feeding Management of Sheep, Goat and Rabbit for sustainable production. (February, 10-12). Organised by *Indian Society of Sheep and Goat Production and Central Sheep and Wool Research Institute,* Avikanagar-304501, Rajasthan. Pp.316-317.

Mathur, B. K., Mathur, A.C., Patel, A.K. and Kaushish, S.K. 1999. Importance of mineral supplementation in buffaloe production in arid region. In proceedings National Seminar on "Sustainble development of buffaloes for milk, meat and draft" 14-16 October 1999. National Dairy Research Institute Karnal-132001. No.5, pp. 163.

Mathur, B.K. (2002). Feeding management of livestock during drought in Arid Zone. DEN NEWS (Desert Environment News) 6(3-4):13-16.

Mathur, B.K. 2002. मरु क्षेत्रा में अकाल के समय पशु पोषण प्रबन्धन (In Hindi). Krishi Lok : Nov. Dec. pp.4-6.

Mathur, B.K. and Mathur A.C. 2002. अकाल में पशुओं में बिमारियों की रोकथामए वं उपचार। (In Hindi). Maru Krishi Chayanika. January to December. 1-3: 51-55

Mathur, B.K. Patel, A.K. & Kaushish S.K. 2000. Utilization of tumba (*Citrullus colocynthis*) seed cake of desert for goat production Indian Journal of Animal Science 70 (4): 431-433.

Mathur, B.K., Kumar Suresh, Ahuja, Usha Rani, Saha, D.K., Friedel, Margaret and Narain, Pratap. 2002. Impact of Availability and Type of Grazing Resources on Livestock health in Arid Zone. Paper presented in *"National Symposium on Grassland and Fodder Research in the New Millennium*. October 30-November 1 2002. Abstracts pp: 157-158; at IGFRI, Jhansi-284003.

Mathur, B.K., Patel AK, Mathur, AC, Abichandani, KK and Kaushish SK, 2000. Wheat straw (untreated treated) consumption by tharparkar heifers during summer under stall fed conditions in "Third Biennial Conference of Animal Nutrition Association" on "Livestock feeding strategies in the new millennium", Nov. 7-9, 2000. CCSHAU., Hissar. No. 282.p.206.

Mathur, BK, Shiv Prasad and Mittal JP, 1989. Effect of tumba cake (*Citrullus colocynllus*) feeding on cattle production in arid region. Indian Journal of Animal Science 59:1464-65.

Mathur B.K, Bhati T.K and Tewari J.C. 2006. Comparative palatability of Prosopis cineraria v/s *Colophospermum mopane* leaves in lactating Marwari goat in arid region. *In:* National Symposium on *"Livelihood Security and Diversified Farming Systems in Arid Region"*. January 14-16, 2006. Organised by Arid Zone Research Association of India, CAZRI, Jodhpur. No. 3-2, pp. 51-52.

Mathur. B.K. 2003. Livestock: human need for sustainability in arid environment. In : *Human Impact on Desert Environment*. (Eds. Pratap Narain, S. Kathju, Amal Kar, M.P. Singh and Praveen- Kumar). Arid Zone Research Association of India and Scientific Publishers (India), Jodhpur, India. pp. 506-514.

Mathur, B.K., Singh J.P., Beniwal R.K. and Singh N.P. 2007. Utilization of salty shrub-Lani (*Salsola baryosma*) of arid region as drought feed for goats. FR-55. In: *International Tropical Animal Nutrition Conference*, October 4-7, 2007 NDRI, Karnal. pp.33.

Mathur, B.K., Patil N.V., Mathur, A.C., Bohra H.C., Bohra R.C. and Sharma K.L. 2009. Comparative Mineral Status of Jodhpur District Villagers' Cattle v/s Institute Farm Managed Tharparkar Cattle in Hot Arid Zone. Proceedings of *"Animal Nutrition World Conference"*, 14-17 Feb., NASC., New Delhi, India. pp.61.

Mudgal VD and Pradhan K. 1989. Proceedings of a consultation on "Non-conventional feed resources and fibrons agricultural residues - strategies for expanded utilization" March 21-29, CCHAU, Hisar, p.139.

Narain, P., Sharma, K.D., Rao, A.S., Singh, D.V., Mathur, B. K. and Ahuja, Usha Rani 2000. *Strategy to combat Drought and Famine in the Indian Arid Zone.* Central Arid Zone Research Institute, Jodhpur, 65 p.

Panda M.R. and Baidya S. (2004). Chemotherapeutic trial of Zycloz bolus against endo-paracitics infection in naturally infested grazing cattle. *Indian veterinary journal.* 81:922-924.

Patil N.V, Mathur B.K, Bohra H.C and Patel A.K. 2005. Relative preference index for arid forages for goat and sheep. National Seminar on *"Conservation, Processing and Utilization of Monsoon Herbage for Augmenting Animal Production"*, Deceember 17-18, 2005. Organized by Indian Society for Sheep and Goat Production and Utilization, at RRS, CSWRI, Bikaner. pp. 190.

Pratap Narain, Satish Kaushish, Bhati T.K., Mathur B.K. and Patidar, M. 2002. मरु क्षेत्रा में सुखा काल में उपयुक्त पशु प्रबन्धन। IVLP Pamplet No.7 Published by : Central Arid Zone Research Institute, Jodhpur – 342003.

Ramana J V, Krishna N, Parthasarthy M and Prasad JR 1990. effect of urea treatment on the protein degradability of pearl millet straw Indian Journal of Animal Sciences 16 : 1022-1023.

Ranjhan SK 1990. Agro-industrial by products and non-conventional feeds for livestock feeding. Published by ICAR, Krishi Bhawan, New Delhi - 110 001.

Singh Kiran and Schiere J.B., 1995 (Eds). Handbook far straw feeding system. Principles and applications with emphasis on Indian Livestock production. ICAR, New Delhi, India.

2

Feeding of Buffaloes in arid and semi arid regions

B.S. Punia and P.C. Lailer

Central Institute for Research on Buffaloes, Hisar 125 001

India has 3.81 lakh km^2 land under arid and 9.56 lakh km^2 under semi arid conditions. Regions facing annual water deficit of two third or more of the Potential Evapo-Transpiration (PET) are classified as arid and the regions with deficit ranging from one third to two third of the Potential Evapo-Transpiration (PET) are classified as semi arid region (Ramana Rao *et al.*, 1983). Over 60% of arid region of India is in western Rajasthan. In arid region soils are sandy, poor in nutrients with low water holding capacity and prone to erosion by wind and water. Natural vegetation in such climatic conditions is sparse and stunted, predominantly spiny belonging mainly to grass cover type *Dichanthium-Cenchrus-Lasiurus* type and very small area having *Sehima Dichanthium* type (Dabadghao and Shankarnarayan, 1973). The semi arid regions are defined as transition zones between arid and sub humid belts. Almost all precipitation in arid and semi arid regions occurs as rain, except in higher altitude where dew and frost is formed due to great differences between day and night temperatures. Both transpiration and evaporation are high because abundant heat energy is supplied to change the limited amounts of liquid water into water vapour either directly or through biological process thus maintaining the heat balance of the area.

Climate and nutrition of animals: The major part of the Indian sub-continent falls under tropics, where the temperatures are extreme, varying from 2°C to 43°C. There are four well defined seasons namely summer (hot and dry), rainy season (hot and humid), autumn (mild cold and mild humid) and winter (cold and dry). The effect of climate on the livestock is both direct and indirect.

Direct effects of extreme of climate on livestock: The direct effect of climate is both on the vegetative growth, which provides feed supply, and on the animal, which affects production. In India the climatic effect on the vegetation is more marked as availability of fodder for livestock is affected.

Indirect effect of extreme of climate on livestock: The major indirect effect of climate is on the quantity and quality of feed available for ruminants. The most important climatic factors which affect the plant growth and hence the quantity of feed available are temperature, length of daylight, intensity of solar radiation and rainfall.

Arid and semi arid lands experience stress due to climatic variability and increasing requirement of livestock and human population. Over exploitation of natural resources by increasing biotic pressure happening is throughout the country but in arid and semi arid areas the situation is more desperate. The population pressure gets further compounded by droughts causing extreme pressure on vegetation. The existing grazing pressure in arid area is very high i.e. 1 to 4 ACU/ha against the normal 0.2 to 0.5 ACU/ha (Beniwal and Singh, 2005). It has resulted in alarming scarcity of feed and fodder, which poses great problems in meeting nutritional needs of livestock.

An area of about one million hectare in semi arid India is subjected for grazing 7.4 million cattle, 5.6 million sheep, 3.38 million goats and 6.6 million other animals. Over grazing has not only reduced the production of biomass over the years but also deserted considerable area. Use of improved management practices, protection of native pasture and introduction of improved varieties of legumes and grasses suited to respective agro-climatic zones may cope up the effect of over grazing upto certain extent (Singh *et al.*, 2005).

In the arid and semi arid regions of India where rainfall is low and erratic, the availability of grazing is seasonal. During rain good pastures are available whereas during the dry season dry and poor quality herbage is available. The drinking water for the animals also becomes a big problem in the arid and sub-arid regions. Table 1 shows the common feeds and fodders of arid zone and their composition.

Table 1. Common feeds and fodders of arid zone and their composition (on % DM Basis)

Feed and Fodder	DM	CP	CF	Total Carbohydrate
Legume straw				
Moth (*Phaseolus aconitifolius*) Chara	87.4	9.7	25.2	73.6
Gram (*Cicer arietinum*) Straw	91.3	4.8	44.6	85.6
Groundnut (*Arachis hypogea*) Straw	89.1	8.1	40.63	80.5
Guar (*Cyamopsis tetragonoloba*) Phalgati	87.3	8.9	24.1	75.7
Average	88.8	7.9	33.6	78.8
Cereal Straw				
Sewan grass (*Lasiurus sindicus*)	88.0	3.1	36.8	84.4
Bakaria (*Indigofera cordifolia*)	82.2	6.2	28.2	73.1
Anjan/Dhaman (*Cenchrus ciliaris*)	83.9	8.2	30.54	73.8
Lampa (*Aristida depressa*)	86.9	2.1	36.8	81.1
Gokhru (*Tribulus terrestris*)	83.7	19.11	16.75	62.1
Dub (*Cynodon dactylon*)	88.2	10.47	28.17	75.4
Gramna/Blue panic (*Panicum antidotale*)	91.4	7.26	40.47	83.6
Makra (*Dactyloctenium aegyptiacum*)	87.7	8.32	35.19	75.7
Bui (*Aerva tomentosa*)	87.9	5.0	32.1	82.4
Karad (*Dichanthium annulatum*)	85.3	7.39	28.17	75.8
Grass mixture	88.5	4.39	32.43	84.0
Average	86.7	7.4	31.9	77.7
Shrubs and Trees				
Babool (*Acacia arabica*)	93.2	7.04	33.32	84.5
Phog (*Calligonum polygonoides*)	87.1	6.5	30.2	79.1
Siras/Sares (*Albizzia lebbek*)	85.4	16.81	31.52	67.7
Peepal (*Ficus religiosa*)	84.3	9.02	15.93	68.3
Khejri (*Prosopis cineraria*)	84.1	14.7	15.14	68.1
Jhar beri (*Zizyphus numularia*)	92.4	11.53	33.82	80.6
Kikar (*Acacia nilotica*)	87.7	15.98	7.50	69.7
Kher (*Acacia catechu*)	87.2	13.03	22.55	73.5
Average	87.46	12.03	23.82	72.74

Source: Singh and Saini, 2002

Forage resources in arid and semi arid region: Forage includes browse and herbage which can be consumed as such by the animals or harvested and fed to the animals.

1. Cultivated fodder production: On account of gradual reduction in range and forest land and availability of poor quality and inadequate bio-mass from grazing land, cultivation of fodder crop and production of crop residues has become an integral part of the sustainable livestock productivity in the country. Table 2 shows the cultivated fodder species suitable for different regions. There is ample scope for production of fodder crops under dry land region of the northern plains without affecting the rabi crops (Singh *et al.*, 2005).

Table 2. Cultivated fodder species suitable for different regions in India

S.N.	Regions	Rainfed	Irrigated
1	Arid	Jowar, bajra, moth, guar, lobia,	Lucerene, berseam, oats, maize, bajra.
2	Semi Arid	Jowar, bajra, moth, guar, lobia, velvet bean, field bean, moong	Jowar, maize, lobia, teosinte, lucerene, berseem,sarson, turnips, hybrid napier, oats, sudan grass, guinea grass, setaria, Rhodes.
3	Sub-humid	Dinanath grass, jowar, lobia, rice bean, velvetbean, teosinte, sunnhemp	Berseem, oats, sudan grass, hybridnapier, guar, jowar, maize, para, Rhodes, setaria.
4	Humid	Jowar, Dinanath grass, rice bean	Berseem, oats, hybrid napier, guinea, lucerene, sarson, turnips, setaria, para jowar
5	Lower hills	Jowar, lobia, velvet bean, field bean, guar	Maize, jowar, oat, berseem, Lucerne, hybrid napier, sudan grass, setaria, Rhodes.

Singh *et al.* (2005)

2. Grazing land: Grazing lands in India are considered to be an important source of biomass for sustainable livestock production for the poor and landless farmers. Grazing based livestock production is a predominant occupation and the major source of livelihood for landless and marginal farmers in the country. The area under pasture land is generally unprotected and inhabited by native biomass. The waste lands and lands which are not being used for arable cropping are used for grazing (Beniwal and Singh, 2005).

3. Supplementary feeding in scarcity: Scarcity of feed and fodder resources in arid and semi arid regions and occurrence of frequent famine necessitates huge amount to be spent on transportation of feed from surplus areas. Due to the poor nutritional quality as well as deficient quantity of feeds and fodders in arid and semiarid regions, feeding of supplements (of energy, protein, and minerals) is recommended. In Indian as well as in most of the developing countries, fibrous crop residues are the major feed for buffaloes. These fibrous crop residues are poor source of nitrogen, energy and minerals and can not meet the needs of rumen microbes. Urea molasses mineral block (UMMB) is a very useful to supplement the deficient nutrients like fermentable nitrogen, easily available energy and all the macro and micro essential minerals.

4. Role of buffalo in Indian economy: Buffalo produces 56.0% of total milk production and 30.0% of total meat produced in the country. In 2004-05 buffalo meat export was of value equal to Rs. 1700 Crores and contributions of buffalo milk in Indian economy is Rs 60228.64 crores (Ranjan, 2006). Buffaloes in India are primarily reared for milk production along with limited use of males for draught and meat production. Buffalo contributes significantly to human nutrition.

Animal husbandry in Rajasthan

Table 3 shows the numbers and position of India in world livestock. Table 4 shows the livestock population of the Rajasthan state during the censuses of 1992, 1997 and 2003. This table shows that from 1992 to 2003, there was significant increase in buffalo, swine and poultry populations in the state i.e. stall fed livestock population increased. Camel population is declining at a fast rate during the last three livestock censuses. During 1997-2003 cattle population also decreased. From the data in this table it is evident that farmers are interested in rearing those species of livestock, which are not taken for long distances in search of feed resources and are adopting stall fed livestock rearing. This may be due to several reasons i.e. value addition in agricultural produce to earn livelihood, better transport facilities for transportation of feed, availability of water etc. After the agricultural farming, livestock is the most important source of livelihood in Rajasthan. In western region of the state, which is arid, and the potential for agricultural farming is limited, livestock has provided livelihood security to the farmers and nomadic groups (Chaudhary *et al.,* 1999). Animal husbandry contributes over 15 per cent to the net state domestic product. Rajasthan has the largest livestock population in India; contributing nearly 40% of the wool and 10% milk production in the country (Sinha *et al.,* 2000; Livestock Census, 2004).

Table 3. Livestock population (millions) of important livestock species in India and World (2003)

Species	India	World (FAO estimates)	India's Share %	Rank	Next To
Cattle	185	1366	13.54	First	
Buffaloes	97	167	58.08	First	
Camels	0.6	18.5	3.24	Fifth	Somalia, Sudan, Ethiopia, Pakistan,
Sheep	61	1034	5.9	Third	Australia, China
Goats	124	743	16.7	Second	China
Poultry	489	15854	3.08	Sixth	China, USA, Brazil, Mexico, Indonesia

Source: Agricultural Research Data Book 2004. Agricultural Statistics at a glance, 2003, Directorate of Economics and Statistics, Ministry of Agriculture, Govt of India.

Table 4. Livestock population (thousands) of Rajasthan and percent changes during the last three censuses

Species/Types	1992	1997	% Change	2003	% Change
Cattle	11595	12158	4.9	10854	-10.73
Buffalo	7746	9756	25.9	10414	6.75
Sheep	12168	14312	17.6	16054	12.17
Goat	15062	16936	12.4	16809	-0.75
Horses and ponies	24.63	23.31	-5.3	25.00	7.25
Donkeys	192	186	-3.1	143	-23.4
Camels	730	668	-8.6	498	-25.45
Swine	248	303	22.2	338	11.55
Poultry	3000	4380	46.0	6192	41.36

Source: Statistical Abstract of Rajasthan (2004), Directorate of Economics and Statistics, Govt. of Rajasthan, Jaipur.

Buffalo in arid region

In Rajasthan there are 10414 thousands buffaloes contributing about 60.83% of the total milk production in the state (Tables 5 and 6). The tables also show that except Maharashtra, in all the major milk producing states of the country, more than 50% milk is produced by buffaloes. In arid region, livestock farming makes agricultural farming more sustainable and economically viable. Probably due to this reason the arid region of the country is better ranked than rest of the regions in terms of production of milk (>18%), meat (>20%) and live animals and fibre (> 25%) (Gahlot, 2005). Milk productivity of buffaloes in Rajasthan is lower than in the other major milk producing states in the country because of poor genetic germplasm and nutritional deficiencies. However, farmers of Rajasthan with adjoining Haryana have very good numbers of Murrah buffaloes yielding more than 10 kg. of milk.

Table 5. Share of milk production by cows, buffaloes and goats (000 tones) during 2003-04

States	Cow	Buffaloes	Goat	Total
Gujarat	2062	4116	243	6421
Haryana	1023	4089	109	5521
Madhya Pradesh	2163	2887	338	5388
Maharashtra	3163	2915	301	6379
Punjab	2206	6140	45	8391
Rajasthan	2291	4899	864	8054
U.P	4440	10512	991	15943
A.P.	1750	5209	-	6959
Bihar	1011	1824	345	3180
India	34973	47979	3708	88082

Source: Agricultural Research Data Book 2004. Agricultural Statistics at a glance, 2003, Directorate of Economics and Statistics, Ministry of Agriculture, Govt of India.

Table 6 Share of Milk Production (%) by Cows, Buffaloes and Goats in major milk producing states of India during 2003-04.

States	Cow	Buffaloes	Goat	Total
Gujarat	32.12	64.10	3.78	100
Haryana	19.60	78.31	2;09	100
Madhya Pradesh	40.15	53.58	6.27	100
Maharushtra	49.59	45.70	4.72	100
Punjab	26.29	73.17	0.54	100
Rajasthan	28.45	60.93	10.73	100
U.P.	27.85	65.94	6.22	100

A.P	25.15	74.85	-	100
Bihar	31. 79	57.36	10.85	100
India.	39.71	54.47	5.82	100

Source: Agricultural Research Data Book 2004. Agricultural Statistics at a glance, 2003, Directorate of Economics and Statistics, Ministry of Agriculture, Govt of India.

Microbial activity in rumen: Reticulo-rumen represents a complex ecosystem and efficient fermentation vessel in cattle and buffalo, which is inhabited by bacteria, ciliate, protozoa and anaerobic fungi. These organisms are primarily involved in degradation of carbohydrates, nitrogenous substances and lipid components of the feed and produce fermentation end products, such as, volatile fatty acids (VFA), carbon dioxide, methane and ammonia. The energy generated in the process is used in the biosynthesis of cellular constituents, which then are digested in the lower gut to serve as nutrients for the host. The anaerobic fermentation converts the substrate energy into VF A, which are absorbed and utilized as energy source by the host.

Buffaloes utilize cellulosic fibrous feeds more efficiently than cattle. The superiority has been attributed to the fact that this species posse-sses more and efficient micro flora in the rumen. The distribution of cellulolytic bacteria has been shown to be wider in buffaloes than cattle; however, cellulolytic bacteria differ in their ability to hydrolyse cellulose from different sources of feed in buffalo rumen. Proteolytic and amylolytic bacterial count was recorded 5-7 folds higher in buffalo than in cattle. Feed composition, post-feeding intervals, feeding frequency, soluble sugars, lipids and rumen pH influence rumen protozoal population. The predominant microbial species present in the rumen of cattle and buffalo have been listed in Table 7. Protozoa contribute directly or indirectly to fibre digestion in rumen. Protozoal counts, especially Entodinium, tended to increase in buffalo rumen with an increase in dietary concentrate. Negative effects of lipids on fibre degrading organisms in particular have been reported. The fungal attack weakens the tensil strength of fibre fragment; therefore, more sites are available for bacterial action in the rumen. All the anaerobic rumen fungi have been shown to be cellulolytic in nature besides having proteolytic activity. Bacterial counts recorded in buffaloes caused marked increase in the level of ruminal ammonia (Sharma, 2006).

Table 7. Predominant microbial species reported in rumen of cattle and buffaloes

Microbial species	Cattle	Buffaloes
Cellulolytic	*Ruminicoccus flavefasciens,* *Ruminicoccus albus*	*Ruminicoccus flavefasciens* *Fibrobacter succinogenes*
Proteolytic	*Streptococcus bovis,* *Prevotella ruminicola*	*Streptococcus bovis* *Ruminibactor amylophilus*
Protozoal	*Entodinium simplex,* *Diploplastron Epidinium*	*Entodinium candatum,* *Diplodinium Ostracodinium*

Feed utilization on different planes of nutrition: Buffalo exhibited higher number of rumen amyolytic bacteria as well as cellulolytic bacteria on wheat straw and concentrate diets compared to cattle. Buffaloes have been reported of having higher capacity to digest dietary crude protein and crude fibre than cattle. Higher crude fibre digestion in buffaloes may be due to narrow calorie-protein ratio, which is better suited to proliferation of ruminal cellulolytic microbes. The additional causes for a better conversion of feed in buffaloes may be attributed to longer retention of feed in the digestive tract, favorable rumen conditions for ammonia nitrogen utilization, higher capacity to handle stressful environment and a wide range of grazing preferences. Nutrients digestibility with NPN supplements were also observed better in buffaloes compared to cattle (Sharma, 2006).

Precursors to synthesize milk: Nutritional requirements for the general metabolism and for synthesis of milk constituents are derived from feed consumed by the animals and extracted through blood. The substrates attracted from blood by the lactating mammary gland are glucose, amino acids, fatty acids, mineral and vitamin constituents (Table 8). Acetate and butahydroxy butyric acid are prime substrates for ruminants. The energy source for synthetic process is oxidation of glucose and acetate in citric acid cycle.

Table 8. Blood precursors of the milk constituents in ruminants

S.N.	Milk Constituents	Blood Precursors
1	Water	Water
2	Lactose	Glucose
3	Casein	AA
4	B Lactglobulin	AA
5	a Lactalbumin	AA

6	Immune globulins	Immune globulin
7	Fatty Acids	Acetate, J3hydroxybutyrate, blood lipids
8	Glycerol	Glucose, glycerol from tryglycerides
9	Mineral	Minerals
10	Vitamins	Vitamins

Nutrient Requirements and computation of ration for buffalo: The requirements for milk production may be computed in accordance with the information given in Tables 9 and 10. The nutrient requirements for milk are added to the maintenance requirements to arrive at the total requirements of lactating buffaloes. Example in Table 11 illustrates the computation of ration for a buffalo weighing 450 kg and producing 8 litres of milk per day with a fat content of 6.5 per cent. The maintenance requirements of the buffalo are 248 g DCP and 3.4 kg TDN (Table 9). The requirements for milk production are 636 DCP and 3.64 kg TDN (Table 10). Thus the total requirements are 884 g DCP and 7.04 kg TDN. The feeds available are concentrate mixture (10% DCP, 75% TDN), green berseem and wheat straw. Illustration in Table 11 shows that the nutritional requirements can be met from these feeds as indicated. Similarly, example in Table 12 illustrates the computation of ration for a buffalo weighing 650 kg and producing 12 litres of milk per day with a fat content of 7 per cent. The maintenance requirements of the buffalo are 327 g DCP and 4.4 kg TDN (Table 9). The requirements of milk production are 996 g DCP and 5.76 TDN (Table 10). Thus the total requirements are 1323 g DCP and 10.16 kg TDN. The feeds available are concentrated mixture (16% DCP, 72% TDN), green maize fodder and barley straw.

Illustration in Table 12 shows that these requirements can be met from these feeds as indicated (Punia *et al.*, 1999).

Table 9. Daily nutrient requirements of adult non-producing Buffaloes (Kearl, 1982)

Body Wt.(kg)	Gain (kg)	Dry Matter (kg/day)	DCP (g)	TDN (kg)	Ca (g)	P (g)	Vitamin A (1000 IU
		Producing 4 kg milk containing 7% fat					
		6.3	205	2.8	14	11	15
		7.0	227	3.1	17	13	17
			248	3.4	18	14	19

500	0	8.2	268	3.6	20	15	21
550	0	8.9	288	3.9	21	16	23
600	0	9.5	305	4.2	22	17	26
650	0	10.3	327	4.4	23	18	28
700	0	10.6	346	4.7	25	19	30
750	0	11.0	364	4.9	26	20	32
800	0	11.5	382	5.2	27	21	34

Table 10. Nutrients Required for Production per kg of Milk, to be added to the Maintenance Requirements (Kearl, 1982)

Fat content (%)	DCP (g)	TDN (kg)	Ca (g)	P(g)
4.0	61	0.34	2.7	2.0
5.0	69	0.38	2.9	2.2
6.0	76	0.43	3.1	2.4
7.0	83	0.48	3.3	2.6
8.0	90	0.53	3.5	2.8
9.0	97	0.57	3.7	3.0
10.0	104	0.62	3.9	3.2
11.0	111	0.67	4.1	3.4

Table 11. Computation of ration for a buffalo (450kg body weight) yielding 8 litres of Milk per day with a fat content of 6.5 percent.

Feed	Fresh weight (kg)	DM (kg)	DCP (g)	TDN (kg)
Concentrate mixture	2.0	1.80	180	1.35
Berseem	30.0	5.40	751	3.60
Wheat Straw	5.5	4.95	000	1.98

Table 12 Computation of ration for a buffalo (650 kg body weight) yielding 12 litres of milk per day with a fat content of 7 per cent.

Feed	Fresh weight (Kg)	DM (kg)	DCP (g)	TDN (kg)
Concentrate mixture	6.0	5.40	960	3.90
Maize green	30.0	7.50	360	5.40
Barley straw	2.5	2.25	18	1.06

In feeding systems where buffaloes are well fed to appetite on roughages/pastures following calving, the bulky nature of the feed may limit the actual intake of nutrients. The animals may then use body reserve to meet the production stimulus. For this reason high producing buffaloes may lose weight following calving. Although meals or concentrates are expensive in comparison to forages but it is desirable to feed at least 2-3 kg concentrate per head per day to buffaloes during early lactation to minimize this weight loss. The objective for the first two to three months is to fully feed the buffaloes to appetite so that they are able to produce to a maximum with high peak yield. Adequate feeding over this period will also minimize the weight loss, enable this loss to be regained rapidly and ensure that buffaloes are in a sound condition at mating. Considerable research has been carried out to study the effects of different types of supplements over the critical early part of the lactation especially in cattle. The most important feature is that continued heavy feeding of a supplement that has a lower feed value than the quality pasture/fodder will reduce lactation production because a degree of underfeeding is being imposed. Heavy feeding of hay or silage to buffaloes particularly high yielder during early lactation means that overall level of nutrient has been lowered and underfeeding results. So under these circumstances rather than depressed production for whole lactation, concentrates should be fed to overcome the degree of underfeeding. A useful guide is to supplement with approximately 0.5 kg of concentrate for each kg of hay or silage on dry matter basis. In another way buffaloes with high milk production potential should be fed adequate concentrate to prepare them to produce at the maximum levels in subsequent lactation.

REFERENCES

Agricultural Research Data Book (2004). IASRI Publication, Pusa, New Delhi.

Agricultural statistics at a glance (2003). Directorate of Economics and Statistics, Ministry of Agriculture, Govt. of India.

Beniwal, R.K. and Singh, J.P. (2005). In: Seminar on "Augmentation of lives-stock Production using newer technologies, 11-12 March 2005, College of Veterinary and Animal Science, Bikaner (Rajasthan).

Chaudhary, A.L.; Gahlot, A.K. and Beniwal, B.K. (1999). Paper presented in the 4th Agricultural Science Congress, 14-21 February, B.M. Birla Science and Technology centre, Jaipur (Rajasthan).

Dabadghao, P.M. and Shankarnarayan, K.A. (1973). The grass covers of India, ICAR, New Delhi.

Gahlot, A.K. (2005). In Proceedings Seminar on "Sustainable Livestock Production in Arid Region" 22-23 March, Bikaner.

Livestock Census (2004). Department of Animal Husbandry and Dairying, Ministry of AGriculture, Government of India, New Delhi.

Kearl, L.C. (1982). Nutrient requirement of ruminants in developing countries. International Feed Stuffs Institute, Utah Agricultural Experiment Station, Utah State University, Logan, Utah, USA.

Punia, B.S.; Lailer, P.C and Sethi, R.K. (1999). Feeding Management for higher Production and Reproduction in Buffaloes. CIRB Publication.

Rajhan, S.K. (2006). In: Symposium on "Buffalo for rural upliftment" 27-30 May 2006, Bombay Veterinary College, Parel, Mumbai. Pp. 1-9.

Ramana Rao, B.V.; Rao, G.G.S.N.; Gupta, B.S. and Malakar, A.R. (1983). Influence of commencement of rain on crop production in dry districts of Rajasthan. Mausam 34: 335-338.

Sharma, D.D. (2006). In: Symposium on "Buffalo for rural upliftment" 27-30 May, 2006. Bombay Veterinary College, Parel, Mumbai. Pp. 27-41.

Singh, G.P. and Saini, N. (2002) Role of anaerobic fungi in fibre digestion and its special significance of Camel nutrition. Indian Dairyman, 54: 64-68.

Singh, N.P.; Triphati, P. and Chaudhary, U.B. (2005). Potential Monsoon herbage of northern Plaints In: Seminar on "Conservation, Processing and utilization of Monsoon herbage for Augmenting Animal Production", 17-18 December, 2005. CSWRI Arid Region Campus, Bikaner (Rajasthan). India. pp. 49-53.

Sinha, K. Rajiv; Bhatia, Sonu and Bishnoi Ritu (2000) Desertification control and range land management in the Thar Desert of India. Indira Gandhi Centre for Human Ecology, Environment and Population Studies. Rajasthan University, Jaipur (India) pp. 15.

Statistical Abstract of Rajasthan (2004), Directorate of Economics and Statistics, Govt. of Rajasthan, Jaipur.

3

Advances in balanced feeding of Camel for its optimum draught potential

G.R. Purohit

Ex-Dean, College of Veterinary and Animal Science, R.A.U., Bikaner-334001

The history of development of camel is narration of a series of evolution, migration and extinction of interest as well as concern, which communicate occurrence of early evolution of camelidae entirely in North America alongwith surprising spread in the late tertiary age to the South America and via Asia to Africa, becoming extinct in North America and recognize dry land ecosystem as habitat.

The dromedary was probably domesticated on the south cost of the Arabian Peninsula in the region of present day Yemen and Oman around 3000 to 4000 years ago and then introduced with the spice trade to North Africa and the horn of Africa. Today there are about 15 to 16 million dromedaries in Africa, Middle East and the Indian sub-continent.

The camel production is besieged with rampant insulting nature of adverse climatic conditions and onslaughts of scarcity of feed and water frequented by recurring low rain fall, failure of crops, meager vegetation resulting in famines, subjecting livestock to nutritional crises. But this unhored pseudo-ruminant seems to have accommodated so well to the arid and semi-arid regions and in the socio-economic pattern of the pastoral industry that inspite of all these aforesaid adversities, all the versatilities and adaptability of the camels to the arid and semi-arid zones due to their anatomical as well as behavioral peculiarities, physiological adjustment and foraging ability such as browsing habits, rendered them non-vulnerable to competition to scarce vegetation reinforced with rema-rkable refractoriness to water deprivation in the dry hot arid region.

The outstanding tolerance to the rugged climate of high temperature, water deprivation, endurance for hunger and scarcity has made this animal to be of great use. Besides, the cushioned feet and muscular fore-limbs supporting pull, the lashing hind-limbs bring about a four wheel drive action in sandy soil. These features in combination with peculiar behavioral superiority as keen sense of night navigation rendered this animal since ancient time as most preferrential or even some times only riding and pack animal in the desert.

As far as utility of this animal in arid tract is concerned, it is refered to as an excellent means of carrying load, transportation, agriculture and defence services in conditions where other animals are scared and failure. Beside these, its hide, wool, milk and even meat available as by product, additionally contribute for better economy of the rural areas.

However, it is because of their survivability, elasticity, diversity and rewarding remunerative economic value, the density of the camel population is escalating in the desertic countries like, India with just 0.7 million camels in 1966 showed spectacular rising trend and touched 1.45 million in 1990 (FAO 1990), which is third highest in the world. Most of Indian camel population (about 82.20%) is restricted in the arid and semi-arid areas of three states (Rajasthan, Haryana and Gujarat) with maximum concentrations i.e., 59.30% of total in Rajasthan alone.

If given a chance, camels prefer browsing over grazing mainly due to large size of the neck, conformation of the lips and teeth help to browse and are therefore well adopted as the third story feeders by nibbling tree leaves at the height of 12 to 15 feet, even on the twigs of the trees, which are often spiny. Though camel do not relish grasses but still feed on a wide variety of vegetation growing in nature and donot seem to disfavour any vegetation within reach that are found through out their expense of the desert belt of Asia and Africa and ėat with impunity certain toxic plants such as Subabul (*Leucaena leucocephala*), Cook town iron wood (*Erythrophleum chlorostachys*) and buck bush (*Salsola koli*) etc., responsible for heavy losses in other species (Purohit, 1995).

Initially the camel nutrition research was taken up at Western Regional Animal Nutrition Research Station, Anand, under ICAR project on "Feed and Fodder Survey of Western Region" and at College of Veterinary and Animal Science, Bikaner under ICAR project entitled "Investigation on Camel Nutrition in Rajasthan Desert" and other studies (Mathur and Purohit, 1979a, Gaur, 1979b Gaur, 1982 *et al.*). Under these projects survey, proximate analysis and digestibility/metabolism studies were conducted for common camel feeds, showed that due to shrinking vegetation browsing is not sufficient for obtaining enough nutrients to fulfill the requirements and the natural flora (tree leaves, shrubs, bushes etc.), products of the desert agriculture crops, which are not useful as human food or industries and the by products of leguminous crops such as *moth* straw, *guar phalgati*, gram straw etc., form the principle feed and fodder resources for camel. Recently with the inception of Indira Gandhi Canal the residue of groundnut crop i.e., groundnut fodder, which is very nutritious, have also become an addition in the resources. For working camels supplementation with concentrate such as *Moth, Guar, Gur, Methi* and Alum are generally fed to meet the nutritional needs as well as to upkeep their body. Studies on inclusion of urea as NPN source and protein protection with or without supplementation of urea indicated that protein protection and supplementation of protected protein with urea could be followed safely and effectively (Mathur and Purohit, 1979; Sharma and Purohit, 1992).

It is evident from many nutritional studies that camels are very efficient in utilizing high fibrous diet and the digestibility of crude fibre is comparatively higher in camels than in other ruminants. In the arid zone of Rajasthan the camel browse on many desert vegetation (non conventional) such as *Ker* (*Capparis decidua*), *Neem* (*Azadirachta indica*), *Kikar* (*Acacia nilotica*), pods of *Vilayati- babool* (*Prosopis juliflora*), *Israeli-babool* (*Acacia tortilis*), *Jal* (*Salvadora oleoides*), *Tumba* (*Citrullus colocynthis*), *Kheep* (*Leptadenia pyrotechnica*), *Bui* (*Aerva persica*), *Lani* (*Salsola baryosma*), *Phog* (*Calligonum polygonoides*), *Dhaman* (*Cenchrus ciliaris*), *Bhurat* (*C. biflorus*), blue panic (*Panicum antidotale*), *Murat* (*P. turgidum*), *Bekaria* (*Indigofora cordifolia*), *Gokhru* (*Tribulus terrestris*), *Buara* (*Andropogan juvarancusa*). These fodder vegetations are invariably consumed by the camels during long famine or scarcity period.

The results of some studies showed that there is no effect of high tannic acid contents of feed on digestibility of nutrients particularly crude protein in camel. These animals can vary efficiently utilize the feed plants and even there is no toxic effect of feeding subabool (*Leucaena leucocephala*) which contain appreciable quanties of memosire.

The studies taken up so far illustrated an apparent wide variation in nutrient intake by a relatively uniform population of camels undertaking similar exercise regimes.

The generalized recommendation for nutrient allowances with regards to DCP is particular suitable for raising of young camels with respect to the approximate body weight and to the age could be used in the ration as a thumb rule for practical feeding and formulation of rations (Purohit, 1995) :

Body weight	Age	DCP requirement
350 kg approx.	3.3 Year	0.34 kg
400 kg approx.	3.9 Year	0.47 kg
450 kg approx.	4.6 Year	0.59 kg

Similarly, requirements of metabolizable energy for maintenance (MEn) in adult and pregnant camel have been reported (Jackmola, 1993). These were considered to be lower than that of the cattle and suggested to reduce further during dehydration. The requirement of energy for maintenance and during last quarter of pregnancy is as under :

Body weight (kg)	Adult camel (Mcal/day)	Pregnant camel (Mcal/day)
300	7.21	9.6
400	8.94	11.9
500	10.57	14.1
600	12.12	16.1
700	13.61	18.1
800	15.05	22.6

Additionally, it was observed that dry matter intake and body weight gains were lower in a group of camels receiving no concentrate, than the group receive high and medium quality concentrate indicated that supplementation could be used as a tool for exploiting potential of camel (Jakhmola, 1993). Similarly it was also observed that supplementation of 40% barley as a source of energy was superior and most effective to enhance body weight gain (Sandoo, 1994). Regarding nutrient utilization in pack camels the experiments were carried out at NRCC on Bikaneri camels, with and without stress. It was observed that camels fed *ad lib* dry fodder of *moth chara* (*Phaseolus aconitifolius*) with no water stress after completing safari of 950 km in Thar desert during winter and daily sent to 10 km journey with pack load of 100 kg (2 riders 114 kg) indicated no deficiency problem (Rai *et al.,* 1995). Camels maintained their body weight and could cope up with the work stress. Similarly, camels subjected to water stress by offering water twice weekly, fed same fodder (@ 1.33% DMI proving 0.34 kg DCP and 68.0 mj ME) also did not exhibited any ill effect. The camels could go well with this much work and water stress. It suggests that pack animals could be maintained well on leguminous fodders without essentially supplementing the concentrate.

A few studies have been carried out in our country on utilization of nutrients and nutrient requirements in racing camels. However, recently preliminary studies made in our country at National Research Centre on Camel, Bikaner showed increase in DM and water intake during racing period than at rest, indicated higher nutrient demands of muscular work in racing camel (Nagpal *et al.,* 2000).

A survey was conducted in four large training campuses in the U.A.E. for calculating intake of energy and protein. On comparing these intakes with published recommendations for 450 kg camels performing similar exercise regimes, the majority of rations having barley, green alfaalfa, milk, dates, grass, pellets and mixed grain provide adequate energy and excess protein to meet the needs for maintenance and daily 20 km sub-maximal training exercise.

Based on data published for other ruminants. The daily maintenance requirement were 270 g of DCP and 49 MJ of ME for a 450 kg camel, whereas, it was suggested a MEm of 0.31 MJ MEm/ $kgw^{0.75}$ compared to 0.50 MJ MEm/$kgw^{0.75}$ suggested earlier. It was concluded that camels have a daily maintenance requirement of 145 g DCP/body $wt^{0.75}$ or 1429 DCP for a 450 kg camel at rest.

One of the very important oxygen uptake study on 500 kg racing camel exercised at racing speed of 35 km/h on a treadmill, demonstrated a gross energy use at 100% VO_2 of 0.63 MJ/Min.

So far no information regarding fat requirements for the camel is available. However, the substitution of rumen undegradable fat sources for conventional carbohydrate has been evaluated in racing camels using iso-caloric diets. In this study, it was found that upto 200 gm of protected fat could be substituted for cereal energy sources without metabolic disturbances.

Studies on nutrient intakes of draft camels during rest and load pulling were conducted at NRCC on adult male camels given dry *moth chara* as sole feed. The camels were given pay-load of 2.5 kg/kg body weight by subjecting to two and four wheel carts for 6 h continuously. The findings of the experiment suggested increase in feed intake, digestibility of nutrients and nitrogen retention due to the higher nutrient demand for work. The loss of body weight recorded in load pulling animals indicated the need of nutrient supplemen-tation to maintain the camels' body weight and work output.

Studies on mineral nutrition of camels reported negative calcium and phosphors balance on feeding some feeds and suggested their supplementation. The intake of macro- and micro-elements de-crease with increasing dehydration but their absorption increases.

Further studies on mineral requirements of dry, pregnant, lact-ating and growing camels suggested sodium supplementation in dry, magnesium in pregnant and higher levels of zinc and manganese in lactating and growing baby camels.

From the foregoing accounts, it could be concluded that require-ments suggested for various categories of the camels are still need to be updated. The pack and racing camels could be maintained on nutrient supply from leguminous straws with other insufficient conventional natural flora but the draft camels used for heavy work such as ploughing and pulling need definite supplementation by including concentrates in their diets to keep them in a fit condition. Attempts should be made to formulate least cost balanced rations with inclusion of non-conventional feeds to make it more effective and economical for working camels.

REFERENCES

F.A.O. (1990). Year Book.

Gaur, D.D. (1979). Studies on feed replacement in camels *(Camelus dromedarious)* and formulation of least cost balance ration. M.Sc Thesis, University of Udaipur (Rajasthan).

Gaur, D.D., Purohit, G.R. and Mathur, C.S. (1982). A note on formulation of least cost balance ration for the camels of the Indian desert. *Annals of Arid Zone* 21(4): 263-266

Jakhmola, R.C. (1992). *Annual Report. NRCC*, Bikaner.

Jakhmola, R.C. (1993). Rumen digestion and available feed stuff in Rajasthan. Workshop organized by the Department of Animal Husbandry, GOR, Jaipur.

Mathur, G.N. (1979). Observations on substitution feeds in camels *(Camelus dromedarious)* and economics of feeding. M.Sc. Thesis University of Udaipur.

Mathur, C.S. and Purohit, G.R.. (1979). Interrelationship of nitrogen intake/nitrogen retention and rate of passage of feed in camels. *All India Symposium on protein and non- protein nitrogen utilization in ruminants*, NDRI, Karnal.

Mathur, C.S. and Purohit, G.R.. (1979). Nutritive value of *Bakariya (Indigofera cordifolia)* for camels *(Camelus dromedarious)* 18(4): 269-271.

Purohit G.R.. (1995). Trends in nutrition research on Camels, Horses and Yak. *VII Animal Nutrition Research Workers Conference,* Bombay.

Rai, A.K., Nagpal, A.K. and Khanna, N.D. (1995). Effect of water restriction on nutrient utilization in pack camels under hot-humid conditions. *Indian J. Animal Science* 65: 1256-1262.

Sandoo, N. (1994). Effect of variation in calori intake on the growth performance of camel *(Camelus dromedarious)* calves. *M.VSc Thesis, Rajasthan Agricultural University,* Bikaner.

Sharma, Y.K. and Purohit, G.R. (1992). Effect of protein protection and supplementation on nutrient utilization and performance of camels. *Proceedings of the first international camel conference,* 2ne to 6th Feb., 1992 at Dubai.

4

Supplementary feed blocks and Nutrient Mixture-An approach for balanced feeding during droughts

H.C. Bohra

Central Arid Zone Research Institute, Jodhpur – 342 003

Most of the livestock population in the state is range managed, except during monsoon period, thrives upon dry grasses in the ranges and pastureland, and crop residues in the fallow lands. These crop residues are very low in essential nutrients including fermentable energy, protein, minerals and carotene, which is a precursor of vitamin A. The problem of mineral deficiency is further aggrieves due to high calcium and very low phosphorus content of the crop residues. Calcium-phosphorus imbalances adversely affect the availability of calcium as well as the phosphorus. The dry fodder scarcity is as another problems especially during drought years. Even in the normal rainfall years, the dry fodder and green forages availability, respectively is 30% and 59% of total requirement of the state. This situation is further worsens during drought years. Under such circumstances crop residues like straws and stovers are imported from the neighboring states. The productivity and survivality during scarcity and drought conditions can be ensured by supplementation of essential nutrients to these animals.

There are different means of supplementation of essential nutrients to the livestock, which are fully or mainly thrive upon the roughage diet. The compact feed block is one of the most appropriate means to do so. Beames (1963) in Australia introduced the concept of feeding compact feed blocks, comprised of molasses and urea. Later on this technology was refined and brought to the field with its main objective to improve productivity of the cattle by Leng and Preston (1983). This pioneer work was initiated in India, and now, with the technical assistance of FAO, the technology is in use in more than 30 countries. Sansoucy and Coworkers (1986) had given detailed procedure of production of compact feed blocks. They have also categorized the blocks produced of different formulations. Molasses has been used as a source of fermentable energy. The molasses of above than 85-brix value should be used for making quality blocks. Urea was used as a source of fermentable nitrogen. Its continuous supply, through feed blocks, increased intake of straw by 40% and digestibility by 20% in cattle. The cereal bran is used as an absorbent of the moisture present in the molasses. It also gives structural stability to the blocks. Sansoucy *et al.* (1986) further opined that the blocks could act as an excellent carrier of the common salt, as well as essential minerals. The gelling agents like magnesium oxide, bentonite, calcium oxide, calcium hydroxide and cement were used for compaction of the blocks. The blocks can be made by hot, warm or cold possesses. Garcia and Sansoucy (1995) had described the cold process in details. He opined that the calcium oxide can be replaced by cement, but before use, it should be mixed with 50% water. Quick lime has also been used in block production, but it resulted into harder blocks than blocks comprised of cement. Cement or its by-product like cement kiln dust, used as a mineral supplement in Canada, Italy, USSR and USA, have not shown any adverse effect when fed up to 1% to 3% of the total dry matter, but, USDA has restricted the use of cement kiln dust as it could result in the deposition of heavy metals in animal tissues. Van Houtert and Leng (1986) showed that the addition of small amount of soluble calcium salt of long chain fatty acids (calcium soap) increases efficiency of the use of fibrous feeds.

Sansoucy *et al* (1986) pointed out that the hardness of block affects its rate of intake. High urea blocks also reduce intake. He recorded that the block feeding increased straw consumption. *In sacco* dry matter digestibility in sheep improved from 42.7 to 44.2% in the sheep offered the compact feed-block. Total fatty acids in rumen were increased when lamb consumed the blocks. The increase was more pronounced in propionate and butyrate production. Feeding cake has also resulted in increase in average daily body gain in ruminants. Sansoucy *et al* (1986) opined that the animal become more efficient in using the available nutrients. In Ethiopia it has been reported that by feeding the feed block, the level of feeding concentrates can be reduced by half without adversely affecting the milk production. Recently, El-din *et al.* (1996) have also reported that the blocks can be used to partially substitute concentrate fraction of the diet to maximize utilization of cotton stalk in sheep. Such type of blocks were found to be much useful as scarcity feed supplement, and can be used as survival feed during calamities. Ranjhan (1993) documented that during volcano eruption at Mt Pinatubo in Philippines in 1991-92, more than 800 cattle and buffaloes were given such type of urea-molasses mineral blocks as relief assistance for a period of 1-yearas scarcity feeds.

In India, feed block technology was first introduced by NDDB, which was followed by States' cooperative dairy federations' cattle feed plants. It was followed with development of appropriate formulations of such blocks at HAU, Hisar; PAU, Ludhiana; IVRI, Izatnagar and also at RAU, Udaipur. But all these formulations included one or other inorganic binder and the product was dried by chemical process. In recent past, some work on development of appropriate formulations using locally available feed resources and gadgets, which can be fabricated by local artisans, was initiated. George Kunju and Ram (1986) have conducted extensive studies on development of formulations and production technology of compact feed block, known as urea-molasses mineral blocks (UMMB). They recorded that utilization of crop residues can be augmented by supplementation of source of nitrogen, energy and mineral which increase the microbial protein synthesis in the rumen and indicated that the buffalo feeds can be economize by feeding urea molasses blocks and feeding of concentrate mixture can be curtailed by 40%, with little quantity of cotton seed meal in block supplemented animals. Further, George Kunju and Ram (1989) conducting feeding trial on Surti buffaloes observed that paddy straw consumption was increased in block-supplanted group. On farm trial indicated that there was an appreciable increase in feed consumption, increase in milk fat and above all the block supplementation proved economical at village level nutrient supplementation to the buffaloes. On an average the buffalo consumed daily 350-370 gm block (Table 1) daily. 3 kg block lasted for week's period. In their earlier formulations they incorporated mineral mixture and cottonseed meal, which were replaced by guar meal and phosphoric acid. In their later formulations, they used fertilizer grade di-ammonium phosphate and magnesium oxide. It reduced the cost and improved quality and texture, as well as the palatability of the block (NDDB,

1996-97). Malik *et al* (1996) have studied the effect of feeding block as such or after soaking in water. They found that 400 gm daily block feeding, after soaking in water and than mixing with the wheat straw increased IVDMD of the wheat straw in rumen of the buffalo. The gas production was also high, and opined that UMMB should be fed as *sani*. Bilala and Murdia (1996), studied effect of block supplementation in Holestein-Friesian heifers. They recorded that inclusion of UMMB to 500 gm daily with concentrate feed supplementation increased digestibility of almost all dietary nutrients and resulted into higher body weight gain (62.5 vs 240 gm/day). Daily live weight gain was further increased (292 gm/animal/day) when these animals, in addition to concentrate+block supplementation, offered fishmeal.

Table 1. Percent ingredients and proximate composition of NDDB's and IVRI Urea molasses mineral block

Ingredient	NDDB (%)	IVRI (%)	Composition	Percent
Molasses	45	45	Moisture	3.0
Urea	15	15	Crude Protein	48.0
Mineral mixture	15	8	Ether extract	0.5
Salt	8	5	Crude fibre	1.5
Calcite powder	4	12	Total ash	34.0
Sodium bentonite	3	3	Silica	2.5
Cotton seed meal	10	12	Calcium	8.0
Sodium bentonite	3	-	Phosphorus	1.2

Source: Malik *et al.* (1996); George Kunju and Ram (1986)

Multi-nutrient Feed Block

All these above mentioned formulations used one or other inorganic material as a binder and production process involved drying by exothermic chemical reaction. In these process the gelling of the block was due to the chemical reaction between the soluble salts of calcium and/or magnesium and phosphorus. Some formulations used calcium oxide as gelling agents. In all cases heat is generated, which may cause nutrient losses. The sticking of the feed ingredient during gelling process is the major problem of using active inorganic binder. Bentonite, used as binder, in some of the formulation is non-selective binder of the macro- and micro-minerals, reduces availability of the minerals in the animal gut.

Considering all these points, simple and appropriate formulations (Table 2) and process technology for production of multi-nutrient feed block, using organic binder (Bohra, 2004) and locally available feed ingredients has been developed at CAZRI. Till date 1+12 formulations of compact feed blocks (Fig. 1) have been developed (Table 3). The standard formulation feed block comprised of wheat bran, sugar cane molasses, urea, vitaminized mineral mixture, dolomite, common salt, deoiled soya bean meal, and guar gum dust used as a binder (Table 4 and 5). It was observed that partial or complete replacement of wheat bran in block formulations with other fibrous feed ingredients like pearl barley milling products, malt sprouts, deoiled rice barn, *Bajra* husk and *Ardu* leaves though reduced production cost of the block, but it resulted into low density blocks (Table 3). Pressing of such fibrous ingredient containing mixture required more energy and its upper surface of the final product will be convex instead of flat. Due to low density of such blocks, these are liable to be over consumed by the animals; however, their bulk density can be improved by increasing the level of the binding material. Further evaluation of new feed block formulations indicated that the partial or complete replacement of wheat bran with other fibrous materials reduced the bulk density and subsequently elevated compressive strength of the blocks, however, this rule does not hold true in case of the blocks in which the wheat bran has been fully replaced by chaffed *Cenchrus ciliaris* straw.

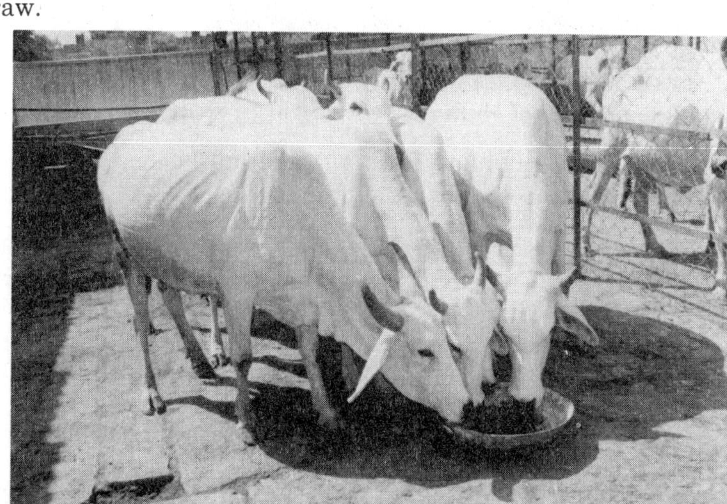

Fig. 1. Tharparker cows licking the supplementary feed block

Table 2. Composition and cost of CAZRI's Multi-nutrient Feed Block and Nutrient mixture

S.N.	Feed Block		Nutrient-mixture	
	Ingredient	Quantity, kg	Ingredient	Quantity, kg
1.	Molasses (44.0%)	5.2	Molasses (46.8%)	5.2
2.	Urea (4.2%) Water (2.1%)	0.5+0.25 lit water	Urea (0.9%) Water (0.45%)	0.1+0.05 lit water
3.	Common salt (4.2%)	0.5	Common salt (4.5%)	0.5
4.	Mineral mixture (4.2%)	0.5**	Mineral Mixture (4.5%)	0.5
5.	Calcite/Dolomite (4.2%)	0.5	Calcite/Dolomite(4.5%)	0.5
6.	Wheat Bran (31.7%)	3.75	Wheat Bran (31.7%)	3.75
7.	*Guar* meal (4.2%)	0.5	*Guar* meal (4.5%)	0.5
8.	Guar gum dust (1.02%)	0.12		
9.	Wrapper	5		

Fig. 2. Ingredients of Multi-nutrient-feed block & -nutrient mixture

Table 3. Dry matter (gm/gm fresh wt) and bulk density of various multi-nutrient feed block formulations

S. No.	Major Constituents		DM, gm/gm fresh wt	Bulk density, gm/ Cu cm	Grading as per bulk density
	Energy sources	Roughages			
Standard	Sugar cane molasses (SCM)	Wheat bran (WB)	0.88	1.25±0.019	B
1	Sugar beet molasses (SBM)	WB	0.84	0.95±0.019	D
2	BM	WB+Cotton seeds, crushed (5:1)	0.83	0.87±0.022	E

3	Feed grade jaggery (J), 72.2% aqueous solution	WB	0.79	1.14±0.010	C
4	J,73.5% aqueous solution	WB	0.78	1.27±0.018	A
5	SCM	Pearl barley, Industry byproduct (BB)+ Guar gum dust (GD), 1.5%	0.86	0.98±0.014	C
6	SCM	BB+GD, 2.5%	0.89	1.15±0.016	C
7	SCM	WB+BB (1:1.1)	0.85	1.01±0.010	C
8	SCM	Malt sprouts	0.85	0.88±0.057	E
9	SCM	Rice polishings (RP)+BB (3.7:1)	0.76	1.23±0.012	B
10	SCM	Rice polishings (RP)+BB (1:1.25)	0.82	1.21±0.019	B
11	SCM	*Ardu* leaves+WB+BB (1:1.8:1.6)	0.78	0.93±0.010	D
12	SCM	*Bajra* husk+RP+ Soyabean meal sol extracted (1:1:0.5)	0.81	0.86±0.024	E
	Mean ± S.E.		0.83± 0.011	1.06±0.042	

Table 4. Feed Block Production Process Technology (Step by Step)

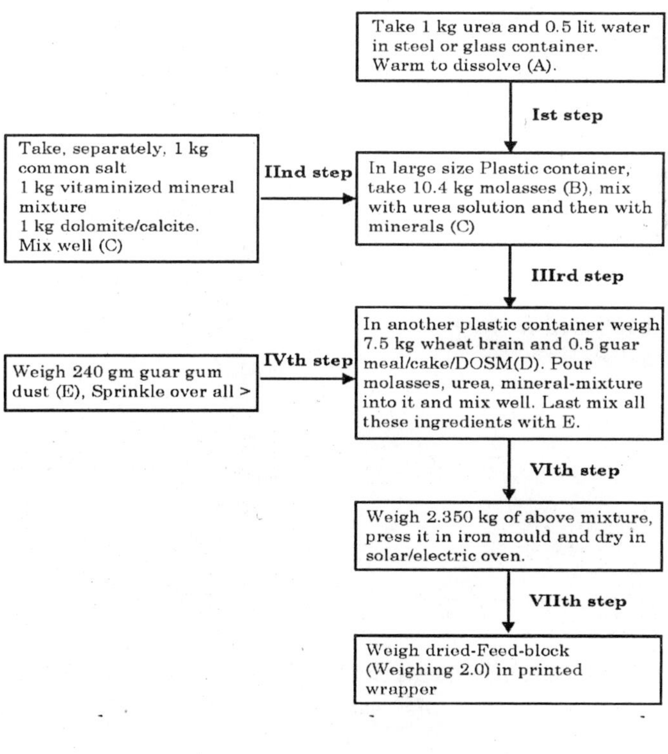

Table 5. Principal and Alternative Ingredients of Multi-nutrient Feed Block and Nutrient-mixture

SNo.	Principal constituent	Alternate constituent (s)
1.	Sugar cane molasses	Sugar beet molasses/ cattle feed grade *jaggery*/ maize strip liquor (maize starch industry by-product)
2.	Urea	-
3.	Common salt	-
4.	Dolomite	Calcite/low silica-dried marble slurry
5.	Mineral mixture containing, amino acids and vitamins (A,D,E)	Mineral mixture containing vitamin A/Mineral mixture containing calcium and phosphorus
6.	Wheat bran	Rice polishings and/or deoiled rice bran, Pearl barley by-products/Malt sprouts/Dried Neem or Ardu leaves, seed free powdered Prosopis juliflora or Acacia tortilis pods/ ground dhaman grass, bajra grain husk and rice polishing mixture/maize gluten feed
7.	De-oiled soyabean meal	Guar meal: korma or churi/any cake of oil bearing seed/ Cotton seed whole, ground
8.	Organic binder (guar gum dust)	Organic: Fenugric seeds, ground Inorganic: Cement, Magnesium oxide, Bentonite or sodium bentonate/ Gypsum/ Calcium oxide

Fig. 3. Feed-block ingredients can be pressed by Screw-type hand-press machine

The physical characteristics and chemical composition of UROMOL (A) produced by the NDDB and MNB (B) produced at CAZRI were compared. The type B block weighed 1.87 kg and NDDB-formulated block produced at RCDF cattle feed plant Jodhpur, weighed 3.02 kg. As such basis, A and B blocks contained 80.4 and 90.0% dry matter, 1.43 and 1.16 gm/cm^3 bulk density, 1.15 and 1.05 gm/cm^3 dry weight density, 15.8 and 100 kg/cm^2 compressive strength, 7.0 and 6.8 pH and 5.19 and 6.37 gm per cent soluble salts, respectively. The compressive strength of CAZRI's block was 6.3 times of the NDDB's uromol block, indicated the superiority of MNB over the UROMOL (Bohra et al., 1999).

Fig. 4. Solar dryer can be used for drying the block

Calcium oxide is being used as binder and gelling agents in several formulations. Experiment conducted to record the effect of incorporated of CaO in standard feed block (ST) formulation @ 25, 50, 75, 100 and 150 gm per block, and pressed blocks were dried in the solar dryer. On dry matter basis these blocks contained, 1.2, 2.4, 3.6, 4.7 and 7.0% CaO. The dry weight of St, and 1-5 blocks was 2110, and 2055, 2090, 2100, 2130 and 2160 gm, volume, 1817 and 1956, 1925, 1829, 2011 and 2227 cm^3, and bulk density was 1.17 and 1.07, 1.10, 1.18, 1.08 and 1.00 gm/cm^3, respectively. The data indicated that the bulk density of block contained 75 gm (3.6%) CaO was highest and it was comparable with the standard feed block. The data further indicated that amongst al the formulations tried, the rate of drying was also fast ($R^2 = 0.9425$) in which 3.6% CaO was incorporated. Calcium oxide incorporation also improved texture of the blocks. In another trial, magnesium oxide was added @ of 0, 4.1, 9.6, 17.5 and 22.75 of total weight of the block. The fresh weight of these blocks were 2.30, 2.45, 2.60, 2.85 and 3.30 kg, and dry weight were 2.00, 2.20, 2.37, 2.60 and 3.1 kg, respectively. Their bulk density was 1.08, 1.22, 1.37, 1.41 and 1.47 gm/cm³, respectively. With increasing the level of MgO, the texture of the blocks improved and colour changed from dark to light gray, the bulk density increased from 1.08 to 1.47 gm/cm³.

Experiment conducted to asses possibility of replacement of molasses, wheat bran and guar meal/deoiled soybean meal in the standard formulation (A) by various maize starch industry by-products, viz, maize-extractives, maize-gluten-feed and meal (B). The A and B blocks weigh 1.89 and 1.80 kg, had 2085 and 1939 cu cm volume and 0.91 and 0.93 gm/cu cm bulk density, respectively. The quality of blocks made of Maize Starch Industry by-products (B) was comparable with the standard block (A), contained molasses and wheat bran except that the former was lighter in colour.

Feeding trial shown higher acceptability of CAZRI's multi-nutrient feed block then the NDDB's UROMOL. The former had showed an increasing feed and water intake in sheep (Fig. 5). The sheep daily consumed 171 gm NDDB's UROMOL (A) and 342 gm CAZRI's MNB (B). The A & B supplemential sheep consumed 740 and 688 gm wheat bran/day, and drank, 2.8 and 3.2 lit water daily. MNB supplemented sheep recorded gain in their live weights, whereas, those on UMB shown reduction. Wheat straw intake was appreciably high (535 gm/day and 2.0 kg/100 kg b.wt) in the sheep supplemented with the MNB than the animal maintained on wheat straw alone (436 gm/day and 1.86 kg/100 kg b.wt). The total dry matter intake including block in former group of animals was 2.6 kg/100 kg b.wt./day. The block supplemented animal daily drank 2.7 lit (10.1 lit/100 kg b.wt), whereas, wheat straw alone fed animals drank 2.3 lit (9.8 lit/100 kg b.wt). During 4 months feeding trial the block-supplemented group recorded 3.6% gain in the live weight, whereas in wheat straw alone fed sheep it was reduced by 8.6%. The average clean wool yield in block-supplemented group was 635 gm, whereas, wheat straw fed seep recorded only 593 gm wool yield (Bohra *et al.*, 2001).

Fig. 5. Sheep licking supplementary block

Rathi heifers, which had free access to the feed block, consumed daily 678 gm block. In another feeding trial, Rathi heifers first offered 300 gm block (B) daily, which was slowly increased to 500 gm/day, and finally to 700 gm/animal/day. Another group was offered wheat straw alone (WS). The WS and B+WS groups of animals consumed daily 2.93 kg and 3.01 kg wheat straw and drank, daily 10.8 and 14.0 lit water, respectively. The average daily dry matter intake, including block in the B+WS group was 3.7 kg/animal/day. The daily dry matter intake, per 100 kg body weight in WS and B+WS group were 1.56 and 2.05 kg, and water intake, 5.6 and 7.8 lit/100 kg b.wt./day, respectively, however, the feed (kg): water (lit) ratio in these animals were not differed significantly; it was 1: 3.7 and 1: 3.8 in WS and B+WS fed animals, respectively, and the digestibility coefficient for the DM in these animals was 44.3 and 48.3%, respectively.

A group of Rathi cows, of an average 313 kg live weight, maintained on *Lasiurus sindicus* pasture and offered daily 1.5 kg RCDF pelleted feed, when had free access to the block, on an average, daily licked 332 gm block. During 5 weeks feeding, the block supplemented group gained 3.5%, whereas, non-supplemental but maintained on grazing and concentrates, recorded 1.9% gain over their initial live weights.

A digestibility trial was conducted to evaluate the effect of feed block supplementation in Rathi cattle. 1 group of cattle had free access to wheat straw (WS) and water, whereas, another group in addition to these offered 700 gm block/animal/day (B+WS). The feed block on dry matter basis, contained 81.7% organic matter, 18.3% total ash, 26.9% crude protein, 0.4% ether extractives, 54.4% total carbohydrates, 15.6% cell wall constituents and 5.1% acid detergent fibre. These values for the wheat straw were 90.0%, 9.1%, 3.1%, 0.4%, 54.4%, 15.6% and 5.1%, respectively. The daily dry matter, crude protein and water intake in WS and B+WS fed groups were 2.93 and 3.96 kg, 97 and 274 gm, and 11 and 14 lit/animal/day, respectively. The digestibility coefficients for dry matter, organic matter, total carbohydrates, crude protein, ether extractives, NDF and ADF in WS and B+WS fed animals were, 44.3 and 48.3; 52.3 and 53.1; 54.4 and 54.0; <0.0 and 43.9; 49.7 and 42.0; 46.5 and 45.8; and 37.5 and 36.7 percent, respectively. The digestibility coefficient for crude protein and DCP value of the wheat straw were <0.0, whereas, of the feed block were 26.85 and 22.4%, respectively (Mondal and Bohra, 2001).

A proto-type solar dryer, which can accommodate 4, 2-kg standard blocks, made of ply wood with outer lining of an aluminum sheet, and inside coated with the black paint were designed and its efficacy to dry the blocks were tried. The maximum temperature of its inner chamber during winter and summer was 65°C and 75°C, respectively.

The drying efficacy of electric and solar dryer was compared. The temperature of non-draft type electric oven varied between 65°C -70°C. It can accommodate 12 blocks and the proto-type solar dryer can accommodate 4 blocks. Its diurnal temperature varies between 26°C-78°C. Under both the conditions, it took 4 days to dry the blocks completely. The electric consumption in drying 12 blocks was 8 units (1.35 Rs/block). In case prototype solar dryer, the cost of installation comes to about Rs. 1000/-, there was no any recurring expenditure. The blocks dried in both the dryers were not differed in quality or texture, but those dried in the electric dryer were dark gray and the solar-dried were giving reddish tinge.

As per the farmers' views the feed blocks formulated and produced at CAZRI were well accepted by the livestock. Feed block supplementation, increased feed and water intake. It corrected the craving of animals for unwanted materials (including geophagia) caused by the pica. Farmers' noticed that the block feeding corrected rumination, and there was an increase in the daily milk yield by 20-25%. It became discernible from 5-6th-10th day of introduction of the block. 1, 2-kg block, in case of a cattle and buffalo lasted for 7 and 5 days, respectively. The farmers of Kalyanpur village reported that block introduction in buffalo increased daily milk yield by 1 lit over its initial production of 3.5 to 4.0 lit/day. In case of a buffalo, an increase in 1.5 lit/day had been recorded (Patel *et al.,* 2003). The blocks were also distributed to the farmers of Kutchch region, even a few urea-free blocks, prepared for horses were also distributed. Both were well accepted by the animals.

Establishment of feed block production unit at farmers' level was found to be economic preposition (Rohilla *et al.,* 2005). The initial cost of establishment of village level production of 20 blocks/ day, including cost of feed ingredients required for production of 100 blocks (Rs.1453/-), essential gadgets and drying oven, comes to Rs. 40,000/-. It excluded the cost of land and infrastructures. The production can be initiated in ventilated, asbestos shade of approx 70'x20', in which 2, 15'x20' areas, situated at both the ends of the shade, can be used for storing the ingredients and finished blocks, and a central space of appox., 40'x20', provided with three phase electric connection for ingredient mixer, block press and draught type electric oven can be used for production purpose. The cost of production of 100 blocks using feed grade jaggery (J) and sugar cane molasses (SCM) comes to Rs. 1460/- and Rs. 1310/-, respectively. Assuming a person can produce daily 20 blocks and if working days in the year are considered to be 300 days, the annual profit on production of (20 x 30 = 600 blocks comes to Rs. 32,400 (Rs. 2700/-) and Rs. 41,400/- (Rs. 3450/- per month) if J & SCM, respectively, are used for block production.

Nutrient-mixture

The feed block though well accepted and have immense use in supplementation of essential nutrients to the livestock of all the categories, but browsers like goats due to nibbling habit did not lick such blocks but try to bite it. For such type of animals formulation and technology for production of nutrient mixture (Fig. 6) have been standardized. It involves mixing of all desired ingredients (Table 2), and then spread over polythene sheet in the sun for drying. Feeding trial conducted on Marwari goat kids and does proved its worth in maintaining their health and productivity (Rohilla *et al.,* 2003; Rohilla *et al.,* 2004) in these animals. Marwari ewes supplemented with nutrient mixture produced more milk, per day/animal and recorded prolonged lactation. Daily weight gain in lambs of nutrient-mixture supplemented ewes were also aoppreciably higher that the lambs of control ewes. (Rohilla *et al.*, 2004; 2005). Extensive study was conducted on multiparous lactating parbatsar goats, maintained by landless goat owners of *Raika* community of village Bassi in Nagaur district. All the animals were let out for grazing and browsing, to meet out their nutritional requirement on village common grazing land and agriculture fallow. The treatment group of animals were offered daily 100 gm nutrient mixture after grazing for the period of 60 days. The peak yield of Parbatsar goats, which attined during 90 to 120 days, extended further for next 50 days in treatment group and total lactation yield was also significantly increased in treatment group by 38 lit per lactation. The cost of 60 days supplementation feeding of nutrient mixture was estimated to be Rs. 48/- and by sale of additional milk produced by treatment group of animals, the farmer earned Rs. 342/-. Nutrient mixture has been found economical as returns were to the tune of about 6 times the cost of the nutrient-mixture (Patel and Bohra, 2006).

Fig. 6. CAZRI's animal feed products

Mineral-block

Formulation and process for production of mineral block (Fig. 7) also was standardized. In formulations 1, 2, 3, and 5, common salt and vitamin-mineral mixture were add in 2:1 ratio, and in formulation No. 4, this ratio was 3:1. In formulation 1 maize flour and guar gum were added @ 24.7 and 1.2%, respectively, and molasses added to facilitate the binding of the ingredients. In No.2 neither molasses nor gum was added but in No.3 gum was added @ 1.6%. In both 3 and 4, 300 ml water was added to facilitate mixing of salt and mineral mixture. In No. 3 and 4, binder was added @ 0.8% level. A 50% *jaggery* solution was also incorporated in No. 3 and 4. In No.5 dolomite was also incorporated. In No. 4 guar gum was added in the last, whereas, in No. 5 it was added before adding dolomite, gum and *jaggery* solution. The fresh and dry weight of No. 1, 2, 3, 4 and 5 formulation blocks were 2.66 and 2.38, 1.77 and 1.63, 1.78 and 1.47, 2.54 and 2.40, and 2.54 and 2.40 kg, respectively. The bulk density of these mineralized salt licks were 1.5, 1.8, 1.3, 1.5 and 1.9 gm/cm^3, respectively. No.2 and 3 were brittle, and No. 4 and 5 due to jaggery were soft and sticky in nature. The block made as per formulation No. 1 was appropriate than rest of the formulations.

Fig. 7. Mineral lick

The technologies developed for production of feed-blocks, nutrient-mixture and salt lick, in present context, have immense potential to augment livestock productivity in drought prone areas and can provide employment in the rural sector of our country. These production technology also provides immense scope to augment utilization of traditional and non-conventional feed resources by the livestock. The CAZRI's supplemental-feed block and its production technologies have been evaluated under Institute village linkage project of National Agricultural Technology Programme. Some of the farmers and unemployed graduates have adopted the technology for commercial production of feed-blocks and nutrient-mixture.

REFERENCES

Beames, R.M. Provision of urea to cattle in a salt/urea/molasses block. *Queensland Journal of Agricultural Sciences 20:* 213-230.

Bilala, M.K. and Murdia, P.C. (1996). Supplementation of urea molasses mineral block to crossbred heifers. *Indian Journal of Dairy Science, 49* (6): 365-370.

Bohra, H.C., Mathur, B.K. and Patel, A.K. (2001). Effect of feeding baked, multi-nutrient feed block on feed and water intake and wool yield in Marwari sheep. *Proceedings of X Animal Nutrition Conference on "Emerging Nutritional Technologies for Sustainable Animal Production and Environment Protection"*, abst., No. 19, p., 11. N.D.R.I., Karnal, November 9-11, 2001, 285 + vii p.

Bohra, H.C. (2004). Use of guar gum industry by-productions in feed-block for livestock. Lecture delivered at *Short course on "Taming Drought, Sustaining Soil Productivity and Diversification of Agriculture through Arid Legumes"*. Jodhpur: Central arid zone Research Institute, December 13-22, 2004.

Bohra, H.C., Mathur, B.K., Patel, A.K., and Kaushish, S.K. (1999). Bulk density of feed block produced by the warm process. *Proceedings of XI Animal Nutrition Conference on "Emerging Trend for Livestock and Poultry Feeding Beyond 2000 AD,"* p., 85-86. Hyderabad: Acharya N.G.Ranga Agric. Univ., Anim. Nutr. Soc., India and ICAR, New Delhi, Dec., 2-4, 1999.

Garcia, L.O. and Restrepo, J.I.R. (1995). Multi-nutrient block handbook. Feed and Agriculture Organization of the United Nations, Rome, 28 p.

George Kunju, P.J. (1989). Urea molasses block- a feed supplement for ruminants. Paper presented at the *International work Shop on Rice Straw and Related Feeds in Ruminant Rations.* Kandy, Sri Lanka. March, 24-28, 1986.

George Kunju, P.J. and Ram, Mangat (1989). Feeding of urea-molasses block for higher milk production in buffaloes. *Proceedings of Indo-Australian Seminar on improving straw utilization by buffaloes for growth and milk production.* (Editors: S.S.Sengar and P. Sikka), p., -220-229, Hisar, CIRB. October, 31-November, 1, 1987.

Kaushish, S.K., Bohra, R.C., Bohra, H.C. and Mathur, A.C. 2006. Effect of Multi-nutrient feed block on Tharparker heifers under desert conditions. *Proceedings of the National Symposium on Livelihood Security and Diversified Farming Systems in Arid region,* (pp., 54-55), Organised by Arid Zone research Association of India, Jodhpur, at CAZRI, Jodhpur, during January 14-16, 2006, 138p.

Leng, R.A and Preston, T.R. (1983). Nutritional strategies for the utilization of agro-industrial by-products by ruminants and extension of the principals and technologies to the small farmers in Asia. In; *Proceedings of the Fifth World Conference on animal Production, Vol., 1:* 310-318.

Mondal, B.C. and Bohra, H.C. 2001. Effect of Multi-nutrient feed block supplementation on feed intake and nutrient utilization in Rathi heifers. *Proceedings of X Animal Nutrition Conference on "Emerging Nutritional Technologies for Sustainable Animal Production and Environment Protection"*, abst. No. 153, p. 95. N.D.R.I., Karnal, November 9-11, 2001, 285 + vii p.

NDDB. (1996-97). National Dairy Development Board, Biotechnology Laboratory, *Animal Nutrition and Feed Technology Annual Report.* Anand, NDDB.

P.P. Rohilla, Mahender Chaudhary, Bohra, H.C., Jangid, B.L. and Singh, Y.V. (2005). CAZRI Technology viable in self-employment: A success story. *Technology Innovation Management and Entrepreneurship, Science Tech Entrepreneur E-Zine,* 6 p.

Patel, A.K., Bohra, H.C. and Bhati, T.K. 2003. On farm trial of multi-nutrient feed block on buffaloes in arid region. *Proceedings of the 4th Asian Buffalo Congress, New Delhi.* Indian Society for buffalo development and Asian buffalo association. New Delhi, Feb., 25-28, 2003.

Patel, A.K. and Bohra, H.C. (2006). Nutrient mixture for enhancing milk yield of Parbatsari goats. Spectrum, ICAR News- a science and technology Newsletter, 12 (1).

Raman Malik, Sengar, S.S. and Mehra U.R. (1996). In-vitro fermentation pattern in buffaloes maintained on UMMB and Sani feeding. *Indian Journal of Animal Nutrition, 13* (4): 244-246.

Ranjhan, S.K. (1993). Animal Nutrition in the Tropics. New Delhi: Vikas Publishing House Pvt. Ltd., 542p.

Rohilla, P.P., Bohra, H.C. and Mathur, B.K. (2004). Effect of supplementation of nutrient mixture on growth of suckling lambs and milk yield of lactating sheep. *Proceedings of the National seminar on "Newer Concepts and Challenges in Veterinary Science and Animal husbandry"*, pp., 238-239. Rajasthan Agricultural University, Bikaner, 31st December 2004-1st January, 2005, 320p.

Rohilla, P.P., Bohra, H.C. and Mathur, B.K. (2005). Effect of supplementation of nutrient mixture on milk yield and live weight gain in lactating ewes and suckling lambs. *Indian Journal of Animal Production and Management.*

Rohilla, P.P., Bohra, H.C. and Patel, A.K. (2003). Effect of supplementation of nutrient feed mixture on growth and milk yield of Marwari goats. *Journal of Eco-physiology* *6*(1-2): 11-14.

Rohilla, P.P., Patel A.K. and Bohra, H.C. (2004). Effect of nutrient feed mixture supplementation on performance of Marwari goats. *Proceedings of the National Seminar on "Opportunities and challenges in Nutrition and Feeding Management of sheep"*, pp, 326. Avikanagar: Indian Society for sheep and goat production and utilization and Central Sheep and Wool Research Institute, February 10-12, 2004.

Sansoucy, R., Aarts, G. and Leng, R.A. (1986). Molasses-urea blocks as a multi-nutrient supplement for ruminants (p., 263-279). In: *sugar cane as Feed, Proceedings of an FAO Expert Consultation* held in Santo Domingo, Dominican Republic from 7-11 July, 1986, 319p.

Van Houtert, M. and Leng, R.A. (1986). *Strategic supplementation to increase the efficiency of utilization of rice straw by ruminants.* University of New England, Armidale, Australia. Mimeograph, 4 p.

5

Drought: Impact on agriculture and livestock, major nutritional strategies to overcome ill effects in livestock

N.V. Patil

Central Arid Zone Research Institute, Jodhpur - 342 003

Drought can be defined as adverse moisture index or adverse water balance, which may be due to generally a prolonged dry spell or insufficient rainfall. If dry weather persists and water supply-related problems increase, the dry period can be called a 'drought'.

In India with 70% of its 1000 million population dependent on agriculture and allied activities though have large water resources and have an annual rainfall ranging from 10,000 mm at Cherapunji, 8500 mm in western ghats to 200-300 mm in parts of western Rajasthan and Gujarat, still find relevance in the old adage that Indian agriculture is a gamble dependent on the monsoon. The country faces typical situation of drought and floods simultaneously. Further the quantum of rainfall received may not be as important as the pattern of rainfall, otherwise Cherapunji recording highest rainfall in the country would not face water scarcity during lean season. Whereas, in the parts of Rajasthan, Gujarat the drought like situation often perpetuate which may be termed as normal situation in the region. Widespread crop failures leading to acute shortages of food and fodder adversely affecting human and livestock, health and nutrition; scarcity of drinking water accentuated by deteriorating ground water quality and declining water tables leading to migration are the major manifestations of droughts and these need be taken care of as a management policy as the droughts have become a recurrent feature.

Drought occurrence and its causes in India

In past century, the country has experienced twenty one large-scale droughts with greater frequencies during the periods 1891-20, 1965-90 and 1997-2000. Large parts of the country perennially reel under recurring drought. Over 68% of India is vulnerable to drought. The 'chronically drought-prone areas' – around 33% – receive less than 750 mm of rainfall, while 35%, classified as 'drought-prone' receive rainfall of 750-1,125 mm. The drought-prone areas of the country are confined to peninsular and western India – primarily arid, semi-arid and sub-humid regions. The 1987 drought was one of the worst droughts of the century, with an overall rainfall deficiency of 19%. It affected 59-60% of the crop area and a population of 285 million. The monsoon of 2000 was the 13th consecutive normal monsoon considering country as a whole, but on a regional basis, this was the third consecutive drought year in areas covered by the states of Rajasthan, Gujarat and Andhra Pradesh.

The El Nino phase of the Southern Oscillations (ENSO) having a direct impact on weak or enhanced summer monsoon which is responsible for 70-90% of annual rainfall, is important reason for drought in India (WMO, 1994). Reckless over-exploitation of surface and groundwater caused sharp decline in groundwater levels, leading to a fall in supply, saline water encroachment and the drying of springs and shallow aquifers. In some regions the decline in water levels has been to the extent of 1-2 metres/year. The declining water levels could lead to a 25% drop in harvests in the near future (Seckler, 1998). The rapid depletion of forest cover is also seen as one of the reasons for water stress and drought. While erratic monsoons and drought are intertwined, a host of other reasons, mostly manmade, – poor irrigation systems, pressure from the increasing industrial use of water aggravate drought or create drought-like situations in the country.

Drought classification

There are a number of classifications and considering Physical aspects the droughts may be clubbed into three or four major groups:

Meteorological drought occurs when the seasonal rainfall received over an area is less than 75% of its long-term average value. If the rainfall deficit is between 26-50%, the drought is classified as 'moderate', and 'severe' if the deficit exceeds 50%.

Agricultural drought occurs when there is insufficient soil moisture to meet the needs of a particular crop at a particular point in time. Deficit rainfall over cropped areas during their growth cycle can destroy crops or lead to poor crop yields. It is typically witnessed after a meteorological drought, but before a hydrological drought.

Hydrological drought is a deficiency in surface and sub-surface water supply. It is measured as stream flows and also as lake, reservoir and groundwater levels.

Degraded pasture land

Management on cultural drought: Action of Government functionaries

The agricultural sector is usually the first to be affected because of its heavy dependence on stored soil water. Soil water can deplete rapidly during extended dry periods. If precipitation deficiencies persist, then people dependent on other sources of water begin to feel the effects of the shortage. Those who rely on groundwater, for instance, are usually the last to be affected.

When the situation returns to normal, and meteorological drought conditions have abated, the 'recovery cycle' follows the same sequence. Soil water reserves are replenished first, followed by stream flows, reservoirs/lakes and groundwater.

The endemic nature of drought in certain parts of the country, notwithstanding, drought management takes the form of crises management, which indicates perhaps the absence or lack of long term planning required to tackle natural calamities. Nevertheless the management strategy for drought relief consists of assessment by revenue authorities of the state governments to measure the gravity of drought situation and an area is declared drought affected when there is damage to more than 50% of standing crop. If the situation is beyond the means of the state govt. to handle, central assistance is sought. After visit of an inter-ministerial team to the affected states and submitting report to the Ministry of Agriculture, the recommendations are put up National Calamity Relief Committee (NCRC) for a final decision.

Funds from calamity relief fund, contributed by central and state governments in the ratio of 3:1, are provided for an action plan drawn by affected States to deal with affected livestock, in inter alia includes sourcing and transportation of water, fodder and opening of fodder depots, cattle camps, release of feed grade grains by the FCI and above all coordination of the various agencies engaged in relief work.

Drought and Agriculture production

The immediate effect of drought is decline in food production. The general pattern of rainfall distribution indicate that south west monsoon is important precipitation source in India

An overwhelming majority of cropped area in India – around 68% – falls within the medium and low rainfall ranges. Large areas are therefore. affected if the southwest monsoon plays truant.

Table 1. Cropped area falling under various ranges of rainfall (percentage distribution)

Range of rainfall	Classification	%
Less than750 mm	Low rainfall	33
750 mm to less than 1,125 Mm	Medium rainfall	35
1,125 mm to 2,000 mm	High rainfall	24
Above 2,000 mm	Very high rainfall	8

Source: Ministry of Agriculture-2002 GOI

Most parts of peninsular, central and northwest India – regions most prone to periodic drought – receive less than 1,000 mm of rainfall. The drought of 1965-67 and 1979-80 affected relatively high-rainfall regions, while the drought of 1972, 1987 and 2002 affected low-rainfall regions, mostly semi-arid and sub-humid regions.

There is a high degree of correlation between food grains production and the monsoon rain pattern-production is higher in years when rainfall is in excess of normal and a dip in years of deficit rainfall.

The drought of 1987 was one of the worst in the century. The overall deficiency in rainfall was 19% as compared to 26% in 1918 and 25% in 1972 being worst years. In the two worst affected states of Rajasthan and Gujarat, the rainfall was less than 50% from normal. In these states, the drought of 1987 was the third or fourth in succession resulting in distress to an unprecedented level. Nearly 54000 villages faced acute drinking water problem. The total area and production in Kharif and rabi food grains reduced by 6.7% and 3%, respectively and it was compensated by increase in area and production under rabi oilseeds by 19.9 and 24%, respectively.

No other drought in the past led to such a drop in food production as the 2002 drought. Food grain production was reduced by about 19% (Table 2).

Table 2. Kharif output in drought years

Deficient rainfall years	Monsoon rainfall (% departure from normal)	Kharif foodgrain production (% decline)
1972-73	24	6.9
1974-75	12	12.9
1979-80	19	19.0
1982-83	14	11.9
1986-87	13	5.9
1987-88	19	7.0
2002-03	19	19.1

The percentage fall of kharif and rabi crop acreage, was about 50%. Production of rice, pulses, commercial crops, oilseeds, Cotton and sugarcane declined by about 7 to 19%. This made it the largest rabi drought in any drought year. The drought of 2002 highlighted the vulnerability of irrigated areas to drought.

Simulation models using soil, plant and water (SPAW) developed to assess impact of drought years on pearl millet yield in Jodhpur was found to be close to observed data and therefore, useful for yield estimation (Rao and Saxton, 1995). Similar approaches are needed to be worked out for all agro climatic regions to plan alternate cropping policy in case of failure of principal crops during drought.

Range Plant response to drought

In the rangelands drought not only reduces forage production but it also lowers forage quality. As the soil moisture becomes less available, the leaf and shoot growth is slowed. Tillering in the grasses is reduced. With the persistence of drought conditions – leaves wilt, fold, become discolored and may eventually die. Even under mild water deficit conditions, there is a reduction in cell wall formation, cell division and protein synthesis. Plants adjust to low soil moisture by shutting down and going dormant. This results in reduction in total amount of above and below ground plant biomass production. During drought periods plants utilize carbohydrates of previous growing season that available in the roots, or crown of the plant. This drain of carbohydrates results in loss of root vigor and fewer basal buds develop for next year's growth. Thus drought not only reduces the forage yield in the drought year but also in subsequent years, even if growing season moisture is adequate.

The plants of arid and semi arid regions have evolved to grow and reproduce when soil moisture is sufficient. If soil moisture is not adequate the plant growth will be reduced despite the peak growing season. However, if there is leaf material present for photosynthesis to occur, energy will be provided to the plants roots and to its developing basal buds for next season's growth.

Grazing Management during drought

Overgrazing in the drought areas is a major factor of quality deterioration of pasture and even in the area which has not been overgrazed the plant material available on the range is extremely fibrous and low in CP which not only limits feed intake but also causes malnutrition, reduces production and even influences survivability.

Good grazing management results in much leaf material, which is essential for taking advantage of any moisture that does occur to produce energy for its roots and basal buds. The less energy the plant is able to provide to its roots the more vulnerable it becomes to the drought and to grazing.

A drought may require that livestock numbers be reduced according to forage supply. Rotational grazing system in drought is to be preferred over continuous grazing as periodic rest helps plants maintain vigor. Concentrating more animals in a single herd is recommended over having several smaller herds as by having more animals in a pasture, the entire pasture will be grazed more uniformly and more use of less preferred plants will be made.

Drought Effect on Livestock

The impact of drought on livestock is manifested in four ways---(i) Mortality, (ii) Loss in productivity, (iii) Health of animal and (iv) Loss in fertility.

The major effect due to reduced availability of feed resources is wide scale mortality in animals. In Rajasthan change in livestock composition since 1951 indicate that sheep and camels followed by cows were most vulnerable animals and they perished in large numbers ranging from 10 to 32%. The situation of various states during drought of 2002-03 is given in Table 3.

Table 3. Impact of drought of 2002 on bovine numbers and fodder production

State	No. Bovines affected (Lakhs)	Fodder deficit (lakh Mts)
Gujarat	54.84	2
Andhra Pradesh	153	61.61
Chattisgarh	17.22	4.2
Karnataka	100	60.09
Madhya Pradesh	192	7.8
Maharashtra	189	241.77
Orissa	117	Surplus
Rajasthan	226	149.31
Tamil Nadu	86	Surplus

Source-Ministry of agriculture, GOI, 2002.

Primary survey in 2002 indicated a loss of cows in large numbers i.e. 25 to 50%. The goats and sheep were relatively less affected compared to large animals. These animals depend on common property resources i.e. grazing lands, wastelands, forests etc. and movement to better endowed areas saves them in severe drought, they move out of state and in worst situation they are sold for slaughter. These animals are reared for ready cash and to meet milk demands of families. However, the health of these animals is affected due to loss of vegetation of their choice. When rain occurs there is a change in type of vegetation for grazing and few animals can not adjust and die (10% of the total number, Rathore, 2005).

Nutritional consequences of drought in the ruminants

Drought has important nutritional consequences in terms of scarcity of both, water and feed on a wider scale and it is necessary to ensure its supply for survivability of livestock. In addition to the amount, the quality of water and feed is also deteriorated in drought periods. There has been little fundamental research on the nutritional consequences of drought for ruminants and other farm species, but it is possible to construct from first principles a satisfactory explanation of metabolic mechanisms invoked for survival in a drought and the reasons why, for example the breeds - Marwari and west African goat, Marwari Sheep and Tharparkar Cattle and even donkeys appear to thrive in conditions that are intolerable for the exotic or crossbred animals.

Body weight loss as a measure of under-nutrition

The first and foremost effect of limiting the nutrients to the livestock due to droughts or feed scarcity situations is loss of body condition. If the animal has good fat reserves it will use these as a source of energy during weight loss. Animal deaths in drought are mainly due to exhaustion of body fat reserves. The tissues of animals dying from starvation for example, contain less than one percent crude fat (Champion, 1971). Critical body weight is the body weight below which mortality rate rises rapidly. Definite threshold values are difficult to give and depend on duration and type of under feeding, as well as on the loss of production that is considered acceptable.

The survival or critical body weight varies with species, strain, breed and age of the animal. It has been reported that cattle will die if weight loss is > 20% of body weight, while sheep and camels can tolerate weight loss up to 30% of the body weight (Leng and Preston, 1987). Cronje (1990) gives critical bodyweight losses of 30 -

40% in cattle and sheep. Obviously, the weight loss that can be tolerated depends on the original weight and condition of the animals, as well as on the physiological status of the animals: i.e. young stock tolerates less weight loss compared to mature animals (Cronje, 1990). An adult animal in good condition can also lose more weight than the same animal in poor condition. Knowledge of critical weight losses for different species is required to optimize feed resource allocation during emergency, but these are not commonly available in India. It would also be useful to use or develop condition scoring techniques to estimate how much weight individual animals are able to lose.

The effects of temporary weight loss can be recovered by compensatory gain (O'Donovan, 1984) especially in growing animals. For milk production, the effects of periodic under nutrition on general health and fertility reproduction are however, more likely to be of a long term nature. Lactation curves are unlikely to recover and calving rates and conceptions may be lowered (Allden, 1970; Robinson, 1990).

Body condition loss and reproduction

In conditions when sufficient nutrients, particularly energy are not available to the cows or sheep/goats, a loss in body condition results causing decrease in milk production and reproductive activity may cease. In pregnant cows the end result is light weight calves and open cows. Fertility of the cows decline when their body condition score (BSC) drops to below 4; especially at calving and if these conditions occur during the breeding season, reductions in pregnancy rate can occur. The overall consequences presented in Table 4 shows the reproduction is impacted most by these deficiencies.

To prevent such undesirable effects, cows either must be provided sufficient nutrients to avoid weight losses and maintain production requirements or they must be relieved totally or partially from these nutritional stressors.

Table 4. Influence of inadequate dietary nutrient intake on reproduction in cattle (Corah, 1988)

Nutrient consumption	Reproductive consequence
Inadequate energy intake	Delayed puberty, suppressed estrus and ovulation, suppressed libido and spermatozoa production
Inadequate protein intake	Suppressed estrus, low conception, fetal resorption, premature parturition, weak offspring

Vitamin A deficiency	Impaired spermatogenesis, anesturs, low conception, abortion, weak offspring, retained placentae
Phosphorus deficiency	Anestrus, irregular estrus
Selenium deficiency	Retained placenta
Copper deficiency	Depressed reproduction, impaired immune system, impaired ovarian function
Zinc deficiency	Reduced spermatogenesis

Considerations for Water during a Drought

The water availability during a drought is essential as water helps to regulate the body temperature, for transport of nutrients, etc. The water requirement is related to factors like heat load, production traits and DMI.

The water requirement of an animal is in the range of 2.5 to 4.5 l/kg DMI. Murphy *et al.* (1983) gave the following equation for lactating cows in early lactation in Western conditions mainly:

Water intake (kg/day) = 15.99 + [(1.58 ± 0.271) * DMI] + [(0.9 ± 0.157) * kg milk/day] + [(0.05 ± 0.023) * sodium intake in g/day] + [(1.2 ±0.106) * minimum daily temperature in °C]

Water Requirement depends upon environmental temperature and physiological needs of animals.

A cattle above 27 deg centigrade atmospheric temperature requires 5.5 litres of water per day. A pregnant cattle requires 9 litres of water per day. A milch cow needs 0.87 litres of water per litre of milk produced in addition to its requirement of 5.5.

In case of sheep water requirement is 3 kg per day, 5 kg per day is needed for pregnant sheep and milk.

For other animals water requirement can be assessed on the basis of above principles.

The ratio of Water Intake: DMI will also rise if feed intake is restricted. Therefore, it may be better to express water requirement as a percentage of body weight.

Extending watering frequencies to 2-3 days have proven to be practical during drought periods for large ruminants when water is scarce. This has the advantage of reducing overall feed and water consumption with possible improved nutritional benefits in terms of increased feed digestibility (Leng, 1986) which may be due to increased retention time of digesta in rumen. Body reserves were unaffected due to extending water up to 3 days as compared to daily offered (Leng, 1986).

The effects of restricted water intake include reduced urine output and reduced feed intake. If the water deprivation is severe, dehydration will occur, combined with protein catabolism and finally a failure of the renal function.

Water requirements of cattle may double during hot weather. If cattle do not have sufficient water, they may refuse to eat, experience lower production, and become sick.

During hot, dry weather one concern about cattle drinking stagnant pond water is that animals can die if the water contains certain species of blue-green algae as these are favorable conditions for its growth. But not all blue green algae are poisonous, and the blue green algae that can generate poisonous toxins do not always do so. It congregates on or near the water surface.

Convulsions, bloody diarrhea, and sudden death characterize in toxication with blue-green algae. Clinical signs also include nervous derangement, staggering, tremors, and severe abdominal pain. If concentrations of blue green algae are suspected, dead animals such as mice, muskrats, birds, snakes, or fish may be present near water source.

Dietary Energy and Protein considerations of livestock during Drought

During drought conditions, dietary energy in the available grazing feeds may be low because dietary fiber content is high and because degradable N content cannot sustain sufficient microbial growth in the rumen and/or hindgut to ensure optimal fermentation of potentially digestible dry matter. Thus even though the feed available may be plenty the animal has difficulty in consuming enough ME for maintenance within the constraints of gut fill. Still the tropical breeds excel in their adaptive capability to low energy due to lower metabolic heat requirement at maintenance. And the lower BMR of *Bos indicus* confers greater tolerance to both heat and drought (Frisch and Vercoe, 1977). However extended drought periods causing low forage production results in livestock to range further to obtain necessary amounts of DM to meet their nutritional needs. This increased travel results in the expenditure of additional energy that may result in a loss in body condition for mature animals and reduced gains in immature stock. Besides producing less forage, drought stricken plants also contain lower çoncentr-ations of crude protein, energy, vitamin A and phosphorus. Thus grazing animals need to consume a greater quality of forage to obtain needed nutrients.

Pastures dormant due to drought conditions are usually deficient in protein. The adaptation of the species like goat, sheep to the drought conditions has a greater ability to concentrate urine than cattle or horses (Schmidt-Nielsen, 1964; McFarlane, 1976). This ability to concentrate urine not only conserves water, it also conserves urea produced by catabolism of body protein in an animal at or below maintenance. This urea can be recycled to the rumen or hind gut and incorporated into microbial protein. Thus an animal best able to concentrate urine is best able to digest low protein, high fiber; dry tropical grasses. The ability of ruminants to recycle the end products of catabolism of body protein through the rumen in order to generate more ME and amino acids in conditions of drought is undoubtedly of far greater evolutionary and practical significance. But the dietary protein supply if continuously remains to be low as in prolonged droughts the it is bound to affect the production and health of the animals.

Strategy of meeting feed requirement during drought

Two feeding strategies, one relates with lack of fodder availability and the other with lack of feed quality can be followed to mitigate effects of drought in livestock.

Feeding strategy during limited feed supply

Feeding strategies during droughts having lack of feed can have one or more of the following objectives:

- to feed animals for **maintenance** at a minimum body weight, i.e. a strategy to ensure survival of animals from the available feed;
- to feed animals to maintain weight above the critical or minimum bodyweight including allowance for weight gain in growing animals. In a long drought this would mean that less animals would survive as the available feed would be used quicker;
- to preferentially feed productive stock, such as pregnant and lactating cows. This is called strategic feeding, it would again mean less feed for other animals.

(I) Reallocation of a given amount of feed in the herd

The first major option to overcome a feed shortage is to adjust the animal production to the feed availability. Thole *et al.* (1993) gave some hypothetical and simplified calculations on the economics of different feeding strategies. The mortality in herd was assumed to be entirely caused totally by starvation from the animals other than productive stock in no strategy intervention case because no pasture was available and no feed was available to dying animals. Three strategies to adjust feeding patterns and reduce mortality rates considered were:

(A) Diversion of nutrients from milk producing animals to all other animals of the entire herd. This would **reduce total milk production** from 290 to 60 litres, greatly reducing farmer's income, but **securing the survival** of more animals

(B) Diversion of nutrients from growing animals and bullocks to save other livestock : If the growing animals were fed sufficient nutrients to gain 400 g/day and bullocks and old cows were offered 80% of their requirement, 10 out of the 29 animals at risk can be saved. In this strategy **milk production is not reduced. And 1/3rdmortality can be reduced.**

(C) Diversion of nutrients from both milk producing and growing animals to the starving animals : In this case farmer does not want to lose so many animals and if he/she cannot afford to cut milk production as drastically as in option A, the strategies can be combined. In this some nutrients would be diverted from milking as well as from growing animals, thereby **saving almost all animals** at the cost of 120 kg milk/day.

A reduction of milk yield in Strategy A results in a decreased daily income, which may be difficult to accept for farmers who do not have sufficient other sources of income or cash reserves. The feeding of animals below their requirements (strategy B) is risky because more animals may die eventually if the calamity continues for a longer time.

The choice of strategy is obviously affected by the nature of feed shortage, whether it is regular and predictable or irregular and unpredictable. Both situations are common in the wide variety of Indian farming systems. Strategy B also results in underfed cattle that are less valuable after the calamity. The extra value or costs of these animals is, however, difficult to assess partly because of compensatory gain or lower fertility rates, especially of the younger animals. Also, the value of animal survival is much higher for predominantly Hindu society of India than in many other countries of the world. The relative need to possess an animal that can pull the plough at the next rainy season further complicates the valuation of mere survival as a form of animal production.

(II) Purchase of feed from surplus location

The second major option to save animals is to purchase feeds from surplus regions. High transport costs are involved particularly in a large country like India where extreme droughts and rains can occur simultaneously. Although the calculation shows that at larger distances, i.e. beyond 200 km. the cheaper transport of concentrates favors the cost of nutrients from grains over dry grass.

But the option of grain feeding however has limited value as it is already scarce under normal circumstances and are not likely to be diverted from human consumption in the calamity that also affects human nutrition. Therefore, feeds available for purchase include agro industrial byproducts such as bagasse, grain milling byproducts and molasses or grasses and straws from the fields.

The cost to save the animals by diverting nutrients from milk production to the animals at risk (Option A) was 230 kg milk per day, but by purchasing the feed the cost is about 27% of option (A) on concentrate feed and 17% per day on unconventional feeds like bagasse (Thole et al., 1993). Therefore, the transport of nutrients (Feeds) happens to be the best option to save the animals in drought. This situation changes considerably if feed is not available nearby and transport costs are incurred.

Grasses from forest areas or crop residues from regions with surpluses are commonly transported to feed animals during droughts. The possibility of transport depends on the nature of the calamity. Transportation of feed is only an option if the drought is regional. It has little or no value if infrastructure is highly damaged (floods) or if the calamity is of a much larger scale (severe droughts that occur throughout most of India). In the case of the 1984/1987 drought in Gujarat, it was observed that transport costs were at times higher than the costs of the feed itself. Here under some attention is given to the options to reduce transport cost for the bulky feeds by densification.

(III) Densification of Feeds

One way to reduce transport costs is to compress or to densify the bulky feeds. Jadai *et al.* (1990) reported that there was 2.25 - 2.70 times increase in bulk density of straw based complete feed. Densities can be increased from 65 - 75 kg/m3 to 100 - 110 kg/m3 by baling or even 300 - 500 kg/m3 by briquetting (Bruhn *et al.*, 1975), which seems very high however. Considering the costs of baling grass by hand and bullock driven presses (pada press) the quantity of feed which can be transported in each truck is more than doubled after baling. The calculations of show that when transport distance exceeds 50 and 150 km the costs of baling respectively with pada and hand press are recovered by reduced transport costs. Extra benefits like reduced storage costs, less spoilage and consequently increased quality of the feed are likely to occur but it depends on the type of straw and the method of densification.

(IV) Other strategies to overcome feed scarcity

(a) Reduction of wastage by chaffing

If straw is insufficient for feeding of all animals, the reduction of wastage by chaffing is an option which saves 15-20% of the the straw offered (Shukla *et al.*, 1988). In Haryana and Punjab states chaffing is common on farms, but it is not widely practised in other areas, probably for nutritional reasons as elaborated by De Wit *et al.* (1993).

(b) Urea Treatment of Straw

The transport of urea or ammonia, and treatment might be cheaper than the purchase and transport of additional concentrate or roughage (Schiere and Nell, 1993). Urea-treated straw saves on concentrate feeding, increases milk yield by 1-2 litres/animals a day, offers better economic returns to the farmers and may help reducing land area required for green fodder production. The process is very simple and involves spraying of urea solution uniformly over the straw and storing it for a specific time period. The process of urea ammoniation of straw is as follows Straw-1000 Kg, Urea, 40 kg and Space requirement is 180 x 150 x 150 cm.

In that case the increased intake of treated straw also helps to improve the health and productivity of animals.

(c) Complete feeds

Complete feeds imply a system of feeding all ingredients including : roughages, processed and mixed uniformly, to be made available at *libitum* to the animals (Sharma and Singhal, 1986). It can be in mash and pelleted form when this product is fed as sole source of nutrients. Pelleting feeds increases voluritary intake by 3-30% but adds to processing cost by 57-130% depending upon the type, percentage and original cost of roughages in the ration (Reddy, 1986). If baling of fibrous feeds is practiced, it can be useful to produce complete feeds for use during droughts, i.e. to add some concentrate ingredients. Biologically, the use of complete feeds with an appropriate balance of roughage and concentrates may lead to better utilization of locally available crop residues, agricultural-byproducts and waste. Complete diets for livestock could benefit rural farmers during periods of feed shortage if the feed and transport costs can be kept low. Many complete feeds using locally available byproducts like bagasse, mixed with tree leaves and other unconventional byproducts have been developed. However, the composition of the complete feed needs to be adjusted to the production level of the animals.

(d) Use of Unconventional Feed Resources

Non conventional feeds are helpful to reduce the deficit of animal feeds as well as to make livestock production more economical and profitable. Some feasibly used unconventional feed sources could be:

Kuvadia (*Cassia tora* Linn.) seeds, Vilayati babul (*Prosopis juliflora* Linn.) pods, Babul (*Acacia nilotica* (L) Del.) pods, seeds and babul seeds chuni, Mahuva (*Madhuca indica* J.F. Gmel.) seed cake, Mahuva flowers, Mango seed kernel (*Mangifera indica* Linn.), Sal seed (*Shorea robusta* Geartn F.) cake, Pilludi {*Salvadora oleoides* Dane (var. *persica* Linn.)} cake (deoiled), Subabul (*Leucaena leucocephala*) seeds, Sea weeds (*Sargassum* spp.), Molasses, Rubber seed cake (*Ficus elastica* Linn.), Raintree pods (*Pithecolobium saman*), Tomato (*Lycoperiscon esculentum* Mill.) waste, Isabgul gola and Isabgul lali (*Plantago ovata* Forsk.), Tapioca starch waste *Manihot esculenta* Crahtz.), Tamarind seed powder (*Tamarindus indica* Linn.), Neem cake (*Azadirachta indica* A. Tues.), Karanj cake (*Pongamia glabra*), Niger seed cake (*Guizotia abyssinia* Cass.), Ambadi cake (*Hibiscus canabinus* Linn.), Kusum cake (*Schleichera oleosa* Willd.), Warai bran (*Panicum miliaceum*), Palm (*Borassus flabellifer* Linn.) male flower, Damaged Apple Waste (*Malus sylvestris*), Jowar gluten and jowar cake (*Sorgum vulgare* Pers.), Banana root bulbs (*Misa paradisiaca* Linn.).

Other unconventioal feeds such as Samadi seeds, Shevari seeds, Pilludi fruits, Cher fruits, Hydrol, Malt waste, Orange waste etc.

(e) Nutrient Supplementation strategy during Drought

Pastures and native range that are dormant due to drought conditions may be low in energy, vitamin A, phosphorus, and protein. Meeting the need for these nutrients is important if herd productivity is to be maintained.

If stocking rate is not reduced as per forage availability, supplemental feeding of the deprived nutrients is necessary to maintain herd productivity and alleviate grazing pressure.

Several options are available for supplying energy to cattle on drought stressed pasture. Hay, grain, and crop processing byproducts can all be used to supply energy to grazing cattle. Low-quality forages can also be ammoniated to increase digestibility and protein content.

As a general rule, up to 0.2% of body weight of supple-mental grain per head per day will not result in large decreases in forage intake and digestion

Provide dry cows with approximately 250 g of supplemental crude protein and lactating cows with 300-400 g of supplemental crude protein per day. This can be fed as approximately 400-550 g of soybean meal for dry cows and 800 to 1000 g of soybean meal for lactating cows. Feed 400-500 g per day of a high protein supplement to dry cows and possibly as much as 700-1000 g to lactating cows to maintain forage intake and efficient use of the forage.

Protein supplementation may be necessary for optimum breeding rates during drought conditions. Protein supplements (cottonseed meal, soybean meal), commercial protein rich concentrate mixtures would be appropriate. Alfalfa hay, sunflower meal, safflower meal, as well as other protein meals may also be used as protein supplements.

For some grains, processing may be necessary for optimum use by cattle. Corn and oats can be fed whole but may be used better if coarsely rolled before feeding. Barley and wheat, however, should be coarsely rolled. Avoid fine grinding and rolling, which results in excess fines and dust. These can result in increased incidence of acidosis. In addition, extremely dusty supplements are unpalatable. However, the producer must weigh the additional costs of processing vs. the value of the grain.

Grain processing co-products such as wheat midds, soybean hulls, and corn gluten feed that contain highly digestible fiber provide energy while alleviating much of the negative impact that grain supplementation has on fiber digestibility. In addition, these byproducts provide protein that may also be limiting in drought stressed forages.

Complete Feed Blocks-Strategy for feeding Livestock during droughts

Photo. Fixed complete feed block making machinery.

General Recommendations about supplementation with minerals and Vitamins

Similar salt and mineral mixture during drought as during normal conditions may be provided. During drought, however, phosphorus supplementation is even more critical. A complete mineral supplement containing 12% calcium, 12% phosphorus, 5 per cent magnesium, 0.4% zinc (4,000 ppm), and 0.2% copper (2,000 ppm) has worked well in many areas.

Lack of vitamin A may become a problem during the fall and winter for cows that grazed drought-affected pastures during the summer. Vitamin A is lacking in forages growing under drought conditions and hay produced from drought-affected forages. Cows should receive vitamin A and D booster shots approximately 30 days before calving if they have not been previously supplemented with vitamins. When using byproduct feedstuffs, make sure that the mineral program is balanced. These feeds are typically high in phosphorus and potentially high in sulfur, which may lead to some mineral imbalances.

f) Supplementary feed blocks/mixtures

In case deficiency of multiple nutrients like energy, protein, mineral and vitamins is predicted, the multinutrient bricks/blocks of molasses, urea, mineral and vitamins are preferred for large animals as a lick and multinutrient mixtures can be fed to small ruminants to offset the effects of deficient nutrients.

Conclusion

Recurrent droughts prevailing in one or the other parts of the country creating quantitative and qualitative deficiency of livestock feeds demand a policy to mitigate the nutritional deficiency effects in livestock which may include —

1. Develop the feed resource base through establishment of feed/fodder banks based on early warning about drought from meteorological Department.
2. Proper pasture grazing management methods to protect the biomass and ensure future production from the grasslands once the moisture and climate condition favors.
3. Formulate nutrient supplementation policy to avoid loss of production and health.

4. Conserve, utilize and enrich the feed resources from agricultural byproducts, forest grasses, tree leaves, bushes etc. as a drought proof measure.

REFERENCES

Allden, W.G. (1970). The effects of nutritional deprivation on the subsequent productivity of sheep and cattle. Nutr. Abstr. Rev. 40: 1167-1184.

Bruhn, H.I., Zimmerman, Z. and Niedermeer, R.P. (1975). Developments in pel Leting forage crops. Agric. Engg. 40: 204.

Champion, (1971). In reading material on course on "*Management of livestock sector during floods and famines*". Organised at NIRD, Hyderabad, 30 November 1989.

Corah, L.R. (1988) Nutrition of beef cows for optimizing reproductive efficiency. *Compendium of Continuing Education* 10 (5):695

Cronje, P.B. (1990). Supplementary feeding in ruminants - a physiological approach. South African J. of Anim. Sci. 20 (3): 110-117. *Dairy India*, 1987.

De Wit, J., Dhaka, J.P. and Subba Rao, A. (1993). Relevance of breeding and management for more or better straw in different farming systems. These proceedings.

Friesch, J.E. and Vercoe, J.E. (1977). Food Intake, eating rate, weight gains, metabolic rate and efficiency of feed utilization in Bos Taurus and Bos indicus crossbred cattle, Anim. Prod. 25:343-358.

Jadai, K.K., Rathee, C.S. and Lohan, O.P. (1990). Effect of compaction of roughage based complete feed on digestibility and rumen Parameters. *Ind. J. of Anim. Nutr.* 7(1): 27-30.

Leng, R.A. (1986). *Drought feeding strategies and practice.* Printed in Austrialia by peel valley printery, 3 Hill street, Tamworth, New South Wales, 2340.

Mc Farlane, M.V. (1976) Water and electrolytes in domestic animals. In *Veterinary Physiology* ed. J.W. Phillips pp.461-539 Wright- Scientechnica, Bristol.

Murphy, M.R., Davis, C.L. and McCoy, G.C. (1983). Factors affecting water consumption by lactating cows in early lactation. *J. Dairy Sci.* 66: 35.

Negi, D.S. (2002). Drought and Livestock Farming. *Indian Dairyman*. 54(8): 29-32

NRC, (1989). Nutrient requirements of Dairy cattle, 6th rev. ed., Washington, D.C.

O Donovan (1984). Compensatory gain in cattle and sheep. Nutr. Abstr. and Rev., series B 54(8): 389-410.

O Donovan, 1984. Compensatory gain in cattle and sheep. Nutr. Abstr. and Rev., series B 54(8): 389-410.

Preston, LR. and Leng, R.A. (1987). Matching ruminant production systems with available resources in the tropics and sub tropics. Penambul books, Armidale 245 pp.

Rao, A.S. and Saxton, K.E. (1995) analysis of soil and water stress for pearl millet in an Indian arid region using the SAAW model. J. Arid Environments. 29:155-67.

Rathore, M.S. (2005) State level analysis of drought policies and impacts in Rajasthan. India.Colombo, Sri Lanka, IWMI,40p. (Working paper 93, Drought series paper no. 6).

Reddy, M.R. (1986). Processing of crop residues for formulation of complete feeds, pp. 129-155. Proceedings of fifth Animal nutrition research workers' conference held at Udaipur from 14-17 July 1986.

Robinson, J.J. (1990). Nutrition in the reproduction of farm animals. Nutr. Res. Rev. 3: 253-276.

Schiere, J.B. and Nell, A.J. (1993). Feeding of urea treated straw in the tropics. 1. A review of its technical principles and economics. *Animal Feed Science and Technology*. 54:89

Schmidt-Nielsen, K. (1964) Desert Animals: Physiological problems of Heat and water. Claredon Press ,Oxford.

Sharma, D.D. and Singhal, K.K. (1986). Efficiency of complete feeding system in ruminants, pp. 166-175. Proceeding of 5th animal nutrition research workers conference held at Udaipur, July 1986.

Shukla, P.C., Talpada, P.M. and Desai, M.C. (1988). Scarcity feeds during drought and other natural calamities. Paper presented at the 15th workshop of AICRP on Agricultural byproducts and symposium on latest research trends in livestock and Poultry nutrition, held at Banaras Hindu University Varanasi, March 1988.

Thole, N.S., Sharma, D.D., De Wit, J. and Singhal K.K. (1993) *Feeding strategies during natural calamities. In Feeding of Ruminants on Fibrous crop residues* Eds Kiran Singh and Schiere, J.B. Indian Council of Agril Research Pp 344-357

WMO (1994) Climatic variability. Agriculture and Forestry Tech. Note No. 196 World meteorological Organization, Geneva, P.77.

6

Complete Feed and Fodder Blocks - A strategic approach to combat feed stress situations

P.C. Lailer

Central Institute for Research on Buffaloes, Hisar 125 001

Efficient utilization of crop residues and agro industrial by products as animal feed assumed importance due to shortage of dry roughages, concentrates and green roughage to an extent of 45, 44 and 39% respectively. Due to cultivation of commercial crops, huge quantities of different type of crop residues are produced every year as a renewable resource. These crops residues are rich in fibre and low in nitrogen and are not palatable. Any processing method that improves the nutritive value of these crop residues will enhance the density of nutrients per unit of feedstuff and go a long way to bridge the gap between availability and requirement of roughages to existing livestock population (Anon., 2001-02).

Processing: More than 30 processing methods have been developed experimentally or practically to improve the intake and digestibility of crop residues. These processes include physical, chemical and biological methods. Most of the methods have been technically successful and have been exhaustively investigated.

Physical methods: Physical methods include chaffing, soaking, grinding, block formulation, pelleting, extrusion, steam under pressure and irradiation. Chaffing is a less severe processing method and aimed at preventing selection of highly nutritive tender parts by animals to reduce the particle size and to prevent feed wastage. Feeding of chaffed or long straw did not affect the digestibility in sheep (Devendra, 1983) but consumption of chaffed straw was higher than long straw (Castillo *et al,* 1982). The bulk density increases markedly by grinding, blocks and pellet making. As reported by Reddy and Reddy (1990), grinding increased the density of cottonseeds hulls by 30% and dried forest grasses by 88%. It was also reported that pelleting of ground material increased the density of maize crops by 54% and that of maize straw by 236%. (Reddy and Reddy, 1990). Similarly feed block also increase straw density. Grinding does not improve the digestibility of fibre fractions of crop residues but some times reduces their digestibility as compared to chaffing. This might be due to faster rate of passage from reticulo-rumen (Greenhalgh and Wainman, 1972).

In order to utilize crop residues in the rations of ruminants, they have to be subject to at least a minimum processing like chaffing/grinding. Economical and balanced feeding of livestock for optimum production is extremely important.

Complete feed concept: The term complete feed is used synonymously with the term total mixed ration (TMR) or complete ration (CR). It is a quantitative mixture of all dietary ingredients blended thoroughly to prevent separation and selection and is formulated to meet the specific nutrient requirements. Interest in complete feed is increasing world over because it has balanced ratio of concentrate to roughage and also nutrients consumed and low feeding costs. It ensures better consumption, avoids refusal of unpalatable nutrients, and reduces eating and rumination time and thus increases resting time. An even intake of feed into rumen leads to less fluctuation in release of ammonia and non-protein nitrogen is more efficiently utilized. Waste materials like baggase, cottonseed hulls, corn silage, alphalpha grass silage, ground hay (Ibrahim & Ingalls, 1971), waste paper, berseem hay and poultry dropping were successfully used in complete feed diet (Kesler *et al.,* 1967).

Complete feed block: As the name indicates it is an animal edible product made by compacting complete feed containing roughage and concentrate in a designed proportion capable to fulfill the nutrient requirement for the targeted animal production system. The complete process is normally mechanical. The complete feed blocks may be square, rectangular or in any designed shape depending upon the type of die used in the pressure applying machine.

Methodology for preparation of complete feed Block: Feed resources i.e. roughage, concentrate or any ingredients to be incorporated are properly mixed in desired proportion in mixer for thorough mixing. The mixed material is then put in the hopper of the feed block making machine. Then the presser in applied after closing the hopper for densification. The automatic mode ensures opening of door for release of complete feed block through piston. Then the blocks are ready to use either for feeding or for storage or transportation.

The formulae for complete diets contain different ingredients like roughages (crop residues, legume hay, straws, grasses etc). Protein supplements (ground nut cake, mustard or any cake or gram, moong, moth or urd churi), energy supplements (rice bran, wheat bran, maize, bajra, molasses etc.), mineral supplement (mineral mixture, salt etc.) and vitamins supplements. The roughage portion can be partly replaced by legume hay if available. These ingredients are then processed into mash, blocks or pellets to make complete feed.

Deficit fodder availability due to scanty rainfall coupled with squeezing and degrading grazing land is major constraints for livestock production in arid and semi arid regions. In normal rainfall years, the productively of forage from grasslands as well as from cultivated areas is high during monsoon months leading to surplus availability of forage during the period. Leaf fodder from top feed species is also high during this period. However the forage supply faces an acute shortage during the period of drought. Therefore production cum-conservation strategy is inevitable in maintaining optimum livestock production throughout the year and particularly in time of drought. In Rajasthan stover's of pear millet and moth bean and guar by products constitutes the major source of fodder for livestock. Hay of annual and perennial grasses and other palatable legumes and forbs etc. is conserved traditionally by the inhabitants and fed to animals. So there is need for densification of voluminous dry fodder and hay for easy and economic transport particularly during the period of drought (Beniwal and Singh 2005).

Factors affecting complete feed formulation: The formulation of complete feed blocks is determined and affected by nutrient density, particle size, forage concentrate ratio, source and level of roughage.

Nutrient density: The nutrient density changes depending upon on the type of feed used in the ration. In dry hay or roughage, nutrient density decrease as it tends to bulk up the ration. As the nutrient density decrease, the ration volume per animal increases.

With the increase in density of energy in ration of Holstein cows, feed intake, milk yield, milk protein and lactose yield showed linear increase (Macleod *et al.*, 1984).

Particle size: Particle size is an important factor to be considered for processing of complete feed diets to maintain the milk fat level. Diets having smaller forage particle size enter rumen at a smaller size after initial chewing and swallowing and therefore leave the rumen faster, resulting in increased fractional turnover rate of ruminal DM (Martz and Belyea, 986; Fahey and Berger 1988). Smaller forage particles spend less time in the rumen for microbial digestion thereby decreasing digestibility, particularly of fibre digestion. The increased intake and improved nutritive value of low quality crop residues after mechanical processing is mainly due to particle size reduction. The optimum particle size for improved intake and increasing the density of feed is crucial in the processing of complete feed diets. Amount and physical form of fibre are important in diet of lactating dairy cows to maintain normal milk fat and chewing activities. Intake of long roughages particularly straw in conventional feeding resulted in reduced milk fat where as complete diets maintained normal butter fat level in Friesian cows (Owen *et al.,* 1969). Milk fat depression (MFD) is due to feeding of ration low in fibre (NDF) either because of restricting forage fibre or lowering forage to concentrate intake in the diet (Woodford *et al.,* 1986). Inclusion of only 20% roughage in complete feed resulted in secretion of milk containing relatively low fat (Emery *et al.,* 1964). NRC (1989) recommended a minimum of 25 to 28% dietary NDF of which 75% should be supplied by forage. Insufficient particle size decrease the ruminal acetate to propionate ratio and pH, which in turn lowers milk fat percentage (Grant *et al.,* 1990b and Woodford *et al.,* 1986). At luminal pH below 6.0, growth of the celluloytic bacteria is depressed, increasing the microbes that produce more propionate and decrease acetate to propionate ratio (Grant *et al.,* 1990a). Reduced forage particle size increases DMI; decrease digestibility and retention time of solids in the rumen (Martz & Belgea 1986, Uden, 1987).

Forage concentrate Ration: Complete ration provides adequate and balanced nutrients in an optimum ratio of concentrate and roughage (Krishna *et al.,* 1977). Moseley *et al.* (1976) found that abrupt increase in concentrate increases the consumption of DM and energy, yield of milk and FCM, percentage of milk protein and ruminal concentration of propionate and total acids and decreases percentage of milk fat, total solids, ruminal pH, concentration of butyrate and acetate: propionate ratio. Increase in forage proportion in complete ration has the opposite affects.

Source and level of roughage: The level and source of roughage in the complete feed is apparently of major importance. Suitably processed fibres crop residues can be successfully used as a sole source of roughage in complete feeds for optimum growth and milk production (Reddy, 1988, Reddy *et al.*, 2003). Characteristics and level of roughages in ruminant ration are crucial for maintaining desirable VF A pattern in the rumen and could influence animal performance (Sudweeks, 1977). Chewing activity influence the rate and amount of saliva secretion, which in turn is influenced by the source of forage (Weltch & Smith, 1970). Dietary NDF is negatively associated with digestibility and DMI and is associated with feed intake depression (Kharusani *et al.*, 1993).

Advantages of complete feed and fodder blocks

Nutrients intake and animal performance: In an experiment (Yadav, 1988), adult buffaloes were fed complete feed blocks made of ungrounded wheat straw (Tl), ground wheat straw (T2), ungrounded paddy straw (T3) and ground paddy straw (T4). The DM intake was more in paddy straw based blocks as compared to the wheat straw based blocks. Grinding of straws improved its DM intake. The digestibility coefficients of all the nutrients were higher in paddy straw than the wheat straw based blocks and the coefficients decreased due to grinding of straws. Jaglan (2002) conducted an experiment on crossbred heifers by feeding complete feed blocks made of untreated wheat straw (Tl) and it was replaced by urea treated paddy straw (T2), mustard straw (T3) and sugarcane baggasse (T4) at 50% level respectively. In each experiment roughage to concentrate ratio was kept 60:40. The DM intake and daily body weight gain were higher while the cost per kg live weight gain was maximum in Tl than the other treatments. Samanta *et al.* (2003) found that there is an increase in dry matter intake in female crossbreed calves (15-18%), buffaloes (20%) and in lactating cows (8%) from block form than its mesh form of diet. The higher dry matter intake on feeding of densified complete diet does not influence the digestibility of various nutrients. Similarly in an experiment (Lailer *et al.*, 2006) on buffalo heifers found a significant (P<O.OI) higher growth rate (613.83 gm/day) in complete feed block fed group as compared to control group (447.50 g/day) in which the

same feed ingredients were fed in the conventional way of feeding. The overall increase in dry matter intake was 17.63% in blocks fed group and the increase in dry matter intake from wheat straw was 27.74% in the block fed group.

Enhance bulk density of roughage: The roughage resources are bulky in nature and characterized by low bulk density. Thus, transportation of these materials during emergency (flood, drought, famine, cyclone, earth quake etc.) is not cost effective and accounts huge exchequer. Therefore, making of blocks of above resources enhances the bulk density into several times, which in turn facilitates transportation at cheaper rates from surplus region to deficit areas. Table 2 shows the natural bulk density and bulk density of compressed common roughages.

Reduction in storage space and transportation cost: Conventionally, roughage materials are stored in heap (field) or in godown, which occupies huge space due to low bulk density. Therefore, storing of roughage for peri-urban dairyman is costly affair as space is a costly commodity in cities. Table 3 shows the saving of storage space after making feed blocks.

Table 1. Fodder supply. Demand and deficit (million tonnes) scenario in India up to 2025.

Year	Supply			Demand			Deficit					
	Green	Dry	Total	Green	Dry	Total	Green	%	Dry	%	Total	%
1995	379	421	800	947	526	1473	568	60.0	105	20.0	673	45.7
2000	385	428	813	988	549	1537	603	61.0	121	22.0	724	47.1
2005	390	443	833	1025	569	1594	635	62.0	126	22.1	761	47.7
2010	395	451	846	1061	589	1650	666	62.8	138	23.4	804	48.7
2015	401	466	867	1097	609	1706	696	63.4	143	23.4	839	49.2
2020	406	473	879	1134	630	1764	728	64.2	157	24.9	885	50.2
2025	411	488	899	1170	650	1820	759	64.9	162	24.9	921	50.6

Source: Patil *et al. (2005)*

Table 2. Bulk density of common roughages

S.N.	Roughage Resources	Bulk Density (kg/Cubic meter)	
		As such	Blocks at 4000psi
1	Natural grasses	25.70	453.51
2	Wheat straw	93.99	421.51
3	Rice straw	24.75	395.5 1
4	Urd straw	59.32	543.12

5	Groundnut haulm	56.18	582.67
6	Stylo straw	28.26	485.91
7	Sorghum stover	42.40	559.89
8	Leucaena leaves	83.04	745.50

Source: Samanta et al (2003)

Table 3. Saving of storage space by making feed roughages blocks

S.N.	Roughages	Stored as Such (cubic meter/quintal)	Stored in Block Form (cubic meter/quintal)	% Reduction in space/quintal
1	Natural grasses	3.89	0.22	94.34
2	Wheat straw	1.06	0.24	77.35
3	Rice straw	4.04	0.25	93.80
4	Urd straw	1.69	0.18	89.34
5	Groundnut	1.78	0.17	90.44
6	Stylo straw	3.54	0.21	94.06
7.	Sorghum	2.36	0.18	92.37
8.	Leucaena leaves	1.20	0.13	89.16

Source: Samanta et al. (2003)

Keeping quality: Storing of roughages in an open area sometimes affected by storm or heavy winds, which causes loss of precious biomass. Heavier panicle may settle in storing space but lighter leafy nutritious pan may fly away which reduces the nutritive value of roughages. Similarly the sweeter ingredients of concentrate mixture are eaten by the rat, cockroach etc. which adversely affect the nutritive value of concentrate. The block making does not disintegrate the valuable nutrients either roughage or concentrate and keeping quality remain as such for longer period.

REFERENCES

Anon (2001-02) Annual report, National Institute of animal Nutrition and Physiology (NIANP), Bangalore.

Beniwal, RK. and Singh, J.P. (2005). In: Seminar on Conservation, processing and utilization of Monsoon herbage for augmenting animal production", 17-19 December 2005, CSWRI, Arid Region Campus, Bikaner (Raj.) pp.67-74.

Castillo, I.S., Roxas, D.B., Chaver, M.A., Moniogan, V.G. and Ranjan, S.K. (1982). In: the utilization of fibrous agricultural residues as animal feed (Ed: P.T. Doyle), school of Ag. And Forestry, Univ. of Melbourn, Victoria. pp. 74-80.

Devendra, C. (1983). MARDI Research Bull.

Emery R S, Brown L D and Thomas J W. (1964). Comparison of corncobs and hay in ground, restricted roughage rations affecting milk composition. *Journal of Dairy Science* 17: 1322.

Fahey G C and Berger LI. (1988). Carbohydrate nutrition of ruminants. pp 269. In: The Ruminant Animal: Digestive Physiology and Nutrition. D.C. Church, ed. Prentice-Hall, Englewood Cliffs, NJ.

Grant R J, Colenbrander V F and Mertens D.R. (1990a). Milk fat depression in dairy cows: Role of particle size of alfalfa hay. *Journal of Dairy Science* 73: 1823-1833.

Grant R J, Colenbrander V F and Mertens D.R. (1990b). Milk fat depression in dairy cows: Role of silage particle size. *Journal of Dairy Science* 73: 1834-1842.

Greenhalagh, J.F.D. and Wainman, F.W. (1972). The nutritive value of processed roughages for fettering cattle and sheep. Proc. of British Society of Animal Production. pp. 61-72.

Ibrahim E A and Ingalls J R (1971). Effect of various roughage sources in complete feeds on milk production, composition and rumen metabolism of dairy cows. *Canadian Journal of Animal Science* 51:669.

Jaglan, B.S. (2002). Urea Treated Crop residue based complete feed blocks for growing calves. M.V.Sc. Thesis submitted to CCS HAU, Hisar.

Kesler EM, Chander P T and Branding AE. (1967). Dry molasses product using waste paper as a base for a possible feed for cattle. *Journal of Dairy Science* 50: 1994-95.

Khorasani G Rj, Okine EK; Kennelly J J and Helm J H. (1993). Effect of whole crop cereal grain silage substituted for alfalfa grain silage on performance of lactating dairy cows. *Journal of Dairy Science* 76:3536-3546.

Krishna N, Krishnamohan DVG and Raghva Rao E. (1997).Emerging trend in feed processing and their impact on animal productivity. *In Proc:* VIII Animal Nutrition Workers Conference held at TNVASU Chennai India (Dec. 12-14, 1997). Pp. 102-110.

Kumar, Suresh (2005). In: Seminar on Conservation, processing and utilization of Monsoon herbage for augmenting animal production", 17-19 December 2005,CSWRI, Arid Region Campus, Bikaner (Raj) pp. 1-15.

Lailer, P.C., Dahiya, S.S., Lal, D. and Chauhan, T.R (2006). Growth performance of buffalo heifers fed complete feed block. In Symposium n Buffalo for rural upliftment and annual convention of Indian Society for Buffalo Development", 27-30 May, 2006, Bombay Veterinary College, Parel, Mumbai pp 51.

Macleod G K, Grieve D G, McMillan I and Smith G C. (1984). Effect of varying protein and energy densities in complete rations fed to cows in first lactation. *Journal of Dairy Science* 67: 1421-1429.

Martz F A and Belyea R L. (1986). Role of particle size and forage quality in digestion and passage by cattle and sheep. *Journal of Dairy Science* 69: 1996.

Moseley J E, Coppock C E and Lake G B. (1976). Abrupt changes in forage - concentrate ratios of complete feeds fed *ad libitum* to dairy cows. *Journal of Dairy Science* 59(8): 1471.

National Research Council (1989). Nutrient Requirements of Dairy Cattle, 6th Rev. ed. Natl. Acad. Sci. Washington, D.C.

Owen J B, Miller ELand Bridge P S. (1969). Complete diets given ad libtum to dairy cows. The effect of the level of inclusion of milled straw. *Journal of Agric. Sci. Camb.* 72: 351-57.

Reddy, M.R. and Reddy, G.V.N. (1990). Effect of processing on the nutritive value of crop residues. In Proc. of the National Seminar on the Development of strategies for small ruminant production in India by the year 2000 Tirupati, November 13-17, 1990.

Reddy G V N, Wilhelina, P D and Reddy M S. (2003). Effect of differently processed complete diet on performance of Murrah buffaloes. *Indian Journal of Animal Nutrition* 20: 131-135.

Reddy M R. (1988). Complete rations based on fibrous agricultural residues for ruminants. Proc. of a Consultation on "Non-Conventional Feed Resources and Fibrous Agricultural Residues-Strategies for Expanded Utilization" Hisar, March 21-29. pp 94..

Samanta. A.K, Das, M.M., Singh, K.K. and Kundu, S.S. (2003). Complete Feed Block: A new approach for handling and feeding bulky feed resources. *Indian Dairy Man* 55 (5): 57-59.

Uden P. (1987). The Effect of grinding and pelleting hay on digestibility, fermentation rate, digest a passage and rumen and faecal particle size in cows. *Anim. Feed Sci. Tech.* 19: 145.

Welch J G and Smith A M. (1970). Forage quality and rumination time in cattle. *Journal of Dairy Science.*

Yadav, K.K. (1988). Development of complete famine feed blocks for adult cattle. M.Sc. Thesis submitted to CCS HAU, Hisar.

7

Importance of rumen manipulations towards improving production efficiency of animals in drought situation

S.S. Paul

Central Institute for Research on Buffaloes, Sub Campus, Bir Dosanjh, Nabha-147 201, Punjab

Certain parts of India (e.g. Gujarat, Rajasthan, MP and Orissa) faces drought like situation quite frequently due to failure of monsoon. Livestock contribute more than 50% of the GDP in the drought prone areas. The immediate impact of prolonged failure of rain is on the health of the livestock, due to lack of feed, fodder and drinking water; leading to high livestock mortality and morbidity which has widespread ramification on poor farmer's economical condition. Saving animal wealth under such situation is utmost important as it constitute backbone of the livelihood security system poor farmers of the country. Across arid and the semi-arid regions of the world, that natural pasture is dry and of little nutritive value for many months of the year. Many ruminants kept by small farmers on the drought prone areas feed on nothing but unimproved pastures and edible parts of shrubs and trees throughout their lives. In the dry season proper, when the standing grass is mature and dormant, the grazing comprises mostly hard-to-digest fibre with low protein content. In periods of drought, even this material is in short supply. There is huge potential for improvement in animal production in drought affected areas via manipulating the ruminal fermentation using well established principles of rumen microbial ecology and understanding dynamics of substrate fermentation.

Principles of rumen manipulation during drought

Manipulation of rumen fermentations can be considered as an optimization process, whereby optimal conditions are sought by maximization and/or minimization of fermentation processes, depending on factors such as kind and level of feeding and animal production. Most of the available technologies on rumen manipulation are related to grain fed high producing animals as the developed nations are more concerned with grain based feeding system and not with scarcity feeding or poor quality fibrous feeds. Rumen manipulations for high producing animals fed grain-based diet will not be discussed here. Rumen manipulations aimed at maximization of crude fibre degradation, microbial protein synthesis, maximization of degradation of antinutritional factors and minimization of methane production in poor quality fibrous diet will be dealt in ensuing sections.

Altering the composition of ration and feeding method, or processing ration components

Digestion in the rumen is dependent on the activity of the micro-organisms, which need energy, nitrogen (ammonia, peptides, amino acids), minerals. Poor quality roughages have insufficient nitrogen, sugar, starch and mineral content to satisfy microbial needs and therefore these needs to be supplied to optimize cell wall degradation. Salter and Slyter (1974) advocated a minimum value of 5 mg ammonia-N per 100 ml rumen fluid for optimum microbial growth. Durand (1989) recommended a total N requirements of 26 g N/kg OMAD (apparently digestible organic matter) in the total digestive tract. Traditionally low N diets are supplemented with urea to supply N for optimal microbial growth. It has been suggested that microbial growth will be stimulated by adding extra amino acids, peptides and proteins in diets (in form of oil cakes or fish meal) where supply of amino acid N is low (Wallace, 1991). Studies have indicated that digestion of untreated straw is increased by 9-15% when some source of digestible cellulose, hemicellulose or starch supplements (e.g. alkali or ammonia treated straw, grass, fodder, legume fodder or leaves or pods of leguminous trees, legume straw, beet pulp, molasses or starch rich concentrates) are supplemented in small amounts (Juul-Nielsen, 1981; Silva and Ørskov, 1988; Zorrila-Rios *et al.*, 1989). The leaves and fruit of indigenous fodder trees can be used to supplement pasture grazed in the dry season. However, many trees and plants protect themselves against stresses such as drought, insect predators and microbial infections by producing toxic compounds that adversely effect the host animal on their ingestion or the micro-organisms of its rumen. Several deleterious compounds have been identified in forages and fodder trees. Fodder trees, shrubs and other plant material must therefore be examined for such anti-nutritional factors before their use as feed supplements can be recommended. Certain minerals are limiting in metabolism of rumen microbes in straw based diets which need to be supplemented. The suggested rate of inclusion is 1.8 g sulpher/kg OMD, 5 g phosphorus/kg OMD, 1.5-2.5 g magnesium/kg OMD and 0.5-1.0 mg cobalt/kg DM (Komisarczuk-Bony and Durand, 1991).

Mechanical and chemical or biological processing of feeds like chopping, grinding, pelleting, urea or alkali treatment, treatment with white rot fungus, etc. increase digestibility and reduce loss of energy in form of methane from fibrous diets (Wieser and Wenk, 1970; Walli *et al.*, 1987; Sundstøl, 1988; Moss *et al.*, 1993).

Increasing feeding frequency and feeding complete diet (concentrate and roughage together as total mixed ration) also improve rumen environment and digestibility and performance of animals.

Use of probiotics

Some aerobic viable yeasts or fungi (*Saccharomyces* or *Aspergillus* spp.) added in very small amounts stimulate growth of cellulolytic bacteria in rumen by improving rumen condition (pH, anaerobiosis) or by the supply of bacterial growth factors (Dawson et al., 1990). However, response to probiotics in animals are inconsistent and depend on strain and feeding condition and increase in animal performance is only 7-10% (Wallace and Newbold, 1993). The effects of probiotics on methane production has not been studied carefully. Frumholtz *et al.* (1989) found that ruminal protozoal numbers were reduced 45% by treatment with the probiotic. As methanegenic bacteria have been found to be associated with ruminal protozoa, some decrease in methane may have been due to this decrease in protozoan numbers.

Use of enzymes

Earlier it was believed that rumen microbes produce sufficient enzymes and that exogenous enzymes cannot remain active in rumen because of high proteolytic activity. Now it is established that microbial enzymes produced in rumen are insufficient or limiting and exogenous enzymes are quite stable in rumen. Supplementation of fibrolytic enzymes in diet of ruminants fed poor quality fibrous diet has consistently improved digestibility of feed and performance in ruminants and are being used widely (Yang *et al.*, 1999).

Direct manipulation of rumen microbial population

Rumen fermentation can be manipulated directly by altering microbial population through use of chemicals to selectively stimulate, suppress or eliminate particular group of microbes and by introducing digestion-enhancing bacterial or fungal species or specific toxin degrading microbial species from other animals.

a) Methane inhibitors/propionate enhancers and herbal additives

Many chemical compounds e.g. monensin or other ionophores (Moss, 1993), bromochloromethane (Haque and Bhar, 2001) have been found to be effective as additives for ruminants which reduce methane production and increase propionate production. But their cost, availability, potential risk of residues in milk and meat and microbial adaptation in long use restrict their use in practical roughage based diets. Plant extracts (e.g. *Kutaki picorrhiza*) containing plant secondary metabolites such as essential oils, sarsaponins, phenolics, etc. have been shown to reduce methane production, reduce rumen ammonia production, suppress protozoa and increase feed efficiency by stimulating growth of favourable microbes (Ando *et al.*, 2003; Wadhwa and Bakshi, 2006).

b) Defaunation

Removal of rumen protozoa has been shown to be beneficial for animals fed fibrous diets as protozoa contribute to nitrogen inefficiency in rumen because they engulf bacteria, passed out at a slower rate and undergoes lysis in the rumen thus reduces availability of protein to animals (Kamra *et al.*, 2000; Gregorio *et al.*, 2005). Digestibilty of nutrients increases and better growth rate and wool production has been reported in defaunated animals. In some studies lower rumen ammonia level, lower methane production and higher VFA production has also been reported. This is probably because of increased fungal and bacterial growth and reduction of methanogens in absence of protozoa. Different method of defaunation like chemical treatment (e.g. sodium lauryl sulphate), grain feeding and feeding of saponins are employed for defaunation but maintaining animals in defaunated state is difficult and chemical defaunating agents are toxic to animals.

c) Transinoculation

The fastest way to improve rumen function in an animal is to introduce digestion-enhancing bacterial/fungal species from other animals or to selectively increase populations of species that inhabit the rumen only at low levels. The rumen microbial ecosystem is an extremely diverse and competitive environment. Because of this, many researchers were initially skeptical of being able to administer a DFM (direct fed microbial additive) that would have lasting effects in the rumen. Bacteria and fungus from one ruminant species have been experimentally shown to colonize others successfully. There have been several documented cases of improved animal performance when the rumen has been inoculated with selected microbes. For example, the toxicity of mimosine present in the tropical forage *Leucaena leucocephala* and its degradation products 3,4 DHP and 2,3 DHP (goitrogens) in unadapted animals could be prevented by inoculation of specific bacteria *S. jonesii* isolated from adapted animals or rumen liquorfrom adapted animals (Jones and Megarrity, 1986; Allison et al., 1990; Paul et al., 1998). Another example of a successful application for a DFM is in the case of monofluroacetate poisoning in Australia. This compound is found in some Australian plants and can be toxic to ruminants at doses of about 0.3 mg/kg of body weight. Gregg *et al*. (1998) reported that they successfully inserted the gene encoding for fluoroacetate dehalogenase into several strains of *Butyrivibrio fibrisolvens*. When sheep were inoculated with the altered microbes, they showed reduced toxicological symptoms. The detoxification of 3-hydroxy-4 (1H)-pyridone and monoflouroacetate are good evidence that ruminal fermentation can be modified with lasting effects by a DFM.

Administration of an elite anaerobic fungus, isolated from goats, to sheep resulted in increased digestibility and intake of lingocellulosic feeds (Lee et al., 2000). Similarly Paul et al (2004 a,b) reported that dosing of buffaloes with a non-indigenous anaerobic fungi (Piromyces sp. FNG5 isolated from faeces of wild blue bull) having higher cellulolytic activities than those isolated from buffaloes, goat, sheep, spotted deer, black buck and hog deer resulted in several fold (2.5-4.4 times) increase in cellulolytic microbial population and increase in digestibility of fibre/cell wall components (NDF digestibility increased by 15.7% and ADF digestibility increased by 17.7%) and 58% increase in VFA production. Tripathi *et al* (2006) also observed that a superior fungal isolate from wild blue bull when dosed into rumen of buffalo calves increased growth rate and feed efficiency in buffalo calves by 30%. and tannic rich tree leaves (Paul et al., 2006). Gordon *et. al.* (2000) reported that dosing of sheep with a nonindigenous anaerobic fungi (*Piromyces* sp. CS15 isolated from cattle rumen) having higher cellulolytic activities than those normally isolated from sheep resulted in 12 percentage point increase in voluntary dry matter intake. It was observed that three of the six sheep dosed with strain CS15 contained the fungus at 2 months post dosing.

It has been further demonstrated that the cross-inoculation of rumen fluid or microbes from wild to domestic ruminants alleviates tannin toxicity and enhances the productivity of livestock browsing tannin-containing shrubs (Brooker et al., 1994). In an earlier study undertaken by us indicated that an anaerobic fungal isolate obtained from wild blue bull had high tolerance to phenolic monomers and also degrade phenolic monomers (Paul *et al.*, 2003). Such a fungi when introduced into rumen microflora of buffalo increased degradation of hydrolysable as well as condensed tannin (Paul *et al.*, 2006).

d) Introduction of genetically engineered microbes

Rumen ecology may also be modified by altering the function of bacteria using genetic engineering techniques. In the past ten years, many approaches have been taken in attempts to transform rumen bacteria by introducing a foreign gene or genes into the bacteria. Research in sheep conducted at the Institute of Biology and the Institute of Biotechnology, *University of New England*, both in Armidale, Australia (Gregg *et al.*, 1998), has proved that genetic manipulation of rumen bacteria is feasible. Bacteria grown in the laboratory were shown to recolonize the rumen and genetically modified bacteria did so as efficiently as their unmodified counterparts. The altered bacteria neither died out, as some scientists suspected they would, nor affected the health of the sheep. In future use of recombinant rumen microbes may become practical highly useful and safe in developing countries.

Future research needs

Integration of research findings for rumen manipulation for practical application has the potential to improve survival and performance of animals under drought and scarcity considerably. More basic research need to be undertaken for understanding molecular basis for variation in ability in degrading lignocellulose or plant toxcants among microbes of different habitat which will be helpful in developing superior genetically modified or naturally selected superior strains of microbes for more effective rumen manipulation. Future of extensive rumen manipulation lies with success in development of genetically engineered designer microbes or plants, which need extensive research in this direction.

REFERENCES

Allison, M.J., A.C. Hammond, and R.J. Jones. 1990. Detection of rumen bacteria that degrade toxic dihydroxypridine compounds produced from mimosine. *Appl. Exp. Microbiol.* 56:590-594.

Ando, S., Ishida, T. Hosoda, M. and Bayaru, E. 2003. *Livestock Production Sci*, 82:245-248.

Brooker, J.D., O'Donovan, L.A., Skene, I., Clark, K., Blackall, L. and Muslera, P., 1894. *Streptococcus caprinus* sp. nov., a tannin-resistant ruminal bacterium from feral goats. *Letters. Appl. Microbiol.*, 18: 313-318.

Dawson, K.A., Nman, K.E and Boling, J.A. 1990. Effect of microbial supplements containing yeast and lacto-bacilli on roughage fed rumi-nal microbial activities. *Journal of Animal Science*, 68: 3392-3398.

Durand, M. 1989. Conditions for optimizing cellulolytic activity in the rumen. In: *Evaluation of Straws in Ruminant Feeding*, Chenost, M., Reiniger, P. (Eds.), London, UK. Elsevier Applied Science, p 3-19.

Frumholtz, P.P., Newbold, C.J. and Wallace, R.J. 1989. Influence of *Aspergillus oryzae* fermentation extract on the fermentation of a basal ration in the rumen simulation technique (Rusitec). *Journal of Agricultural Science*, Cambridge, 113:169-172.

Gordon, G.L.R. and Phillips, M.W. 1993. Removal of anaerobic fungi from the rumen of sheep by chemical treatment and the effect on feed consumption and *in vivo* fibre digestion *Letters Appl. Micribiol,,* 17: 220-223.

Gregg, K., B. Hamdorf, K. Henderson, J Kopecny, and C. Wong. 1998. Genetically modified ruminal bacteria protect sheep from fluoroacetate poisoning. *Appl. Environ. Microbiol.* 64:3496-3498.

Gregorio, E., Monforte-Brice, C.A., Sandoval-Castro, L/R., Concepci, M and Leal, C. 2005. *Animal Feed Science and Technology*. 123-124: 313-327.

Haque, N. and Bhar, R. 2001. Effect of mode of supplementation of Mahua seed cake with or without bromochloromethane in roughage based diet on methane production in sheep. *In Proceedings X Animal Nutrition Conference*, Rai, S.N. and Singhal, K.K (Eds), NDRI, Karnal, India.

Jones, R.J. and Megarrity. R.G. 1986. Successful transfer of DHP-degrading bacteria from Hawaiian goats to Australian ruminants to overcome the toxicity of Leucaena. *Aust. Vet. J.* 63:259-262.

Juul-Nielsen, J. 1981. Nutritional principles and productive capacity of the Danish straw-mix system for ruminants. In: Maximum Livestock Production from Minimum Land, Jackson, M.G., Dolberg, F., Haque, M., Saadulllah, M. (Eds), Mymensingh, Bangladesh, Bangladesh Agricultural University, p 287-289.

Kamra, D.N., Singh, R., Chaudhary, L.C., Agarwal, N. and Pathak, N.N. 2000. Soapnut as natural defaunating agent:its effect on rumen fermentation and in-sacco degradability of Jowar hay in buffaloes. *Buffalo Journal.* 1: 99-104.

Komisarezuk-Bony, S. and Durand, M. 1991. Effects of minerals on microbial metabolism. In: *Rumen Microbial Metabolism and Ruminant Digestion*, Jouany, J.P. (Ed), Paris:INRA Editions P179-198.

Lee, S. S., Ha, J.K. and Cheng, K.J. 2000. Influence of an anaerobic fungal culture administration on in vivo ruminal fermentation on *in vivo* ruminal fermentation and nutrient digestion. *Anim. Feed Sci. Technol.,* 88: 201-217.

Moss, A.R., Givens, D.L. and Furniss, S. 1993. A comparison of farm-scale methods of sodium hydroxide on the nutritive value of a winter wheat straw. *Animal Feed Science and Technology*, 41:191-212.

Paul, S.S., Atreja,P.P. and Chhabra, A. (1998) *In vitro* degradation of mimosine and dihydroxypyridine in cattle on transinoculation of rumen liquor from *Leucaena leucocephala* fed goats. *Indian J Animal Nutrition* 16(1): 68-72

Paul, S.S., D.N. Kamra, V.R.B. Sastry and N.P. Sahu (2003) Effect of different phenolic monomers on the growth and hydrolytic enzyme activities of a lignocellulolytic anaerobic fungus isolated from wild nil gai (Boselaphus tragocamelus). *Letters in Applied Microbiology*, 36:377-381.

Paul, S.S., D.N. Kamra, V.R.B. Sastry, N.P. Sahu and N. Agarwal 2004a. Effect of anaerobic fungi isolated from wild and domestic ruminants on *in vitro* digestibility, fermentation characteristics and fibrolytic enzyme production on lignocellulosic feeds by mixed rumen microflora of buffalo. *Reproduction Nutrition and Development* 44:1-7.

Paul, S.S., D.N. Kamra, V.R.B. Sastry, N.P. Sahu and N.Agarwal.2004b. Effect of administration of an anaerobic gut fungus isolated from wild blue bull (*Boselaphus tragocamelus*) to buffaloes (*Bubalus bubalis*) on in vivo ruminal fermentation and digestion of nutrients. *Animal Feed Science and Technology* 115:143-157.

Paul, S.S., Devki N. Kamra, V.R.B. Sastry and Narottam P. Sahu. 2006. Effect of addition of an anerobic fungal culture isolated from wild blue bull (*Baselophus tragocamelus*) to rumen fluid of buffalo (*Bubalus bubalis)* on *in vitro* lignocellulose and tannin degradation, fibrolytic enzyme production and fermentation pattern in presence of tannins. *Journal of the Science of Food and Agriculture* (UK) 86(2):258-270.

Salter, L.D. and Slyter, L.L. 1974. Effect of ammonia concentration on rumen microbial protein production in vitro. British Journal of Nutrition, 32:199-208.

Silva, A.T. and Ørskov, E.R. 1988. Fibre degradation in the rumens of animals receiving hay, untreated or ammonia-treated straw. *Animal Feed Science and Technology*, 29, 277-287.

Sundstøl, F. 1988. Improvements of Poor quality forages and roughages. In: *Feed Science*, Ørskov, E.R. (Ed), Amsterdam:Elsevier, p 257-277.

Tripathi, V.K., Sehgal, J.P. Puniya, A.K. and Kishan Singh. 2006. Isolation of elite anaerobic fungi from fresh faeces of wild blue bull. In *Proceedings XII Animal Nutrition Conference on Technological Interventions in Animal Nutrition for Rural Prosperity*, Jan 7-9, 2006 held at Anand, Gujrat p.118.

Vermorel, M. and Jouany, J.P. 1989. Effect of rumen protozoa on energy utilization by wethers of two diets based on ammonia treated straw supplemented or not with maize. *Australian Journal of Animal Science*, 2:475-476.

Wadhwa, M. and Bakshi, M.P.S. 2006. *Bubalus bubalis*, 12: (in press).

Wallace, R.J. 1991. Rumen proteolysis and its control. In:*Rumen Microbial Metabolism and Ruminant Digestion*, Jouany, J.P. (Ed), Paris, INRA Editions, p 131-150.

Wallace, R.J. and Newbold, C.J. 1993. Rumen fermentation and its manipulation. The development of yeast culture as feed additives. In: *Biotechnology in the Feed Industry*, Lyons, T.P. (Ed.) Nicholasville, Kentucky: Alltech Technical Publications. P 173-192.

Walli, T.K., Gupta, B.N. and Rai, S.N. 1987. Feeding trials with urea treated and fungal treated wheat straw. In:Kiran Singh, T.W. Flegel and J.B. Schiere (eds): *Biological, Chemical and Physical Treatment of Fibrous Crop Residues for use as Animal Feed*, ICAr, New Delhi, India pp. 103-111.

Wieser, M.F., Wenk, C. 1970. Effect of plane of nutrition and physical form of ration on energy utilization and rumen fermentation in sheep. In: 5th *Energy Symposium* EAAP 1970, Schurch, A., Wenk, C. (Eds), Vitznau, Switzerland, p 53-56.

Yang, W,G., Beauchemin, K.A. and Rode, L.M. 1999. Effects of an enzyme feed additive on extent of digestion and lactating dairy cows. *J. Dairy Sci.*, 82:291-303.

Zorrilla-Rios, J., Horn, G.W., Phillips, W.A. and McNew, R.W. 1991. Energy and protein supplementation of ammoniated wheat straw diets for growing steers. *Journal of Animal Science*, 69:1809-1819.

8

Feed processing as complete feed block for sustainable animal production in arid and semi arid regions

Tribhuwan Sharma

College of Veterinary and Animal Science, Bikaner 334 001

Natural calamities those directly affects the feed resource availability are drought, flood, earth quake, cyclone, landslide, hailstorm etc., besides, political disturbances such as war, bandh, restrict movement of feed resources and also the availability (Thole *et al.*, 1991). In such situation primary aim should be the saving of the life of productive and growing stock by providing minimum nutritional inputs. In arid and semi arid area suffering from acute shortage of feed and fodder, the drought or famine is the natural calamity becoming almost regular feature and is considered by many to be the most complex but least understood of all natural hazards, affecting more people than any other. Though drought is normal feature and its recurrence is inevitable yet, there remains much confusion with in scientific and policy community about its characteristics. It is precisely this confusion that explains to some extent the lack of progress in drought management in most part of the world (White, 2000).

In arid area, the persistent drought which often leads to complete crop failure and consequently complete crop residue failure in association with extreme degradation of erst while pastures and drastic reduction of common property resources, exposes livestock to a situation where even very poor quality feed resources (in no way expected to meet minimum nutritional requirement of animals) are not available adequately. The situation is particularly more grave with marginal and small farmers have little or no capacity to produce and store animal feeds and that is why during drought they are the first to be affected. Invariably the sheep and goat are disposed for slaughter during such periods, giving at least some monetary benefits but the large ruminants turns out to be a liability non meetable by the farmers, requires immediate intervention by the government and non government agencies. Further it has also been observed that even after the adequate intervention by these agencies the existing genotype or breed suffers great loss in affected pockets. Though during the next good rainfall year, farmers again purchase cattle from surrounding states or areas, the cattle may be of some good breed but there is essential dilution of the existing germplasm which is perhaps irreversible loss.

Keeping the aforesaid facts in view and realizing that the nutritional inadequacy in the arid and semi arid area for livestock under existing circumstances will be a persistent feature, the efforts are required to be proactive rather than reactive to coup with. To make people living in these areas more self reliant by improving economic viability and sustainability of livestock sector in rain fed arid and semi arid areas, a dependable feeding strategy in terms of fodder banking is required to be formulated for improving nutrient availability.

Need of the new technology

National Waste Land Development Board (1991), through a survey recorded that out of 55 micro regions only 12 regions have showed surplus whereas, 43 regions across the country suffer from scarcity of fodder for livestock. Punjab, Haryana and Western Utter Pradesh are the regions facing problems regarding disposal of cereal crop residues and contrary to this the livestock of Rajasthan, Gujarat, West Bengal, Andhra Pradesh and Orissa are striving for these crop residues.

Looking to the present time availability of feed and fodder resources as well as livestock availability in arid and semiarid area, the self sufficiency seems to be far away and dependence upon the import of fodder from surplus area will be there to meet the requirements at least during exigencies. However, it can also be very well appreciated from experience of past where during emergencies of natural calamities shifting of crop residues from surplus and growing areas has been one of the first crisis management steps taken up by the government agencies. But at the same time it is also important to understand that due to low bulk density of fodder long distance transportation is a non viable proposition. In the financial year 2002-03 alone 7406.20 lakh of rupees were spent by the state government of Rajasthan for the transportation of basal roughages like cereal and pulses crop residues, grasses etc. from surplus and growing areas. It has been estimated that the cost of tariff beyond 200 km is equivalent to actual price of roughage (Samanta *et al.*, 2004). Therefore, need of technology that could facilitate safe and economic transportation by enhancing the bulk density of basal roughage sources is there and it has been observed by many workers that densification as such may play a major role in this regard. Enhancement in bulk density of important feeds by compaction can be appreciated from the Table 1.

Table 1. Bulk density of raw feeds and blocks

S.N.	Feed	Bulk Density of Feed (kg/m3)	Bulk Density of Feed block (kg/m3)
1.	Ground nut straw	81.00	553.63
2.	Ardu leaves	51.50	539.06
3.	Mustard straw	51.48	562.21
4.	Gram straw	66.48	471.94
5.	Wheat straw	57.22	347.22
6.	Jowar straw	70.00	437.20
7.	Paddy straw	70.00	279.00
8.	Forest grass	66.33	503.00
9.	Bajra karbi	119.60	511.10
10.	Cenchrus grass	67.50	486.84
11.	Arhar straw	55.00	689.06
12.	Sugarcane bagasse	61.0	567.50
13.	Cluster bean	109.00	596.67
14.	Wheat straw (with molasses 5%)	70.00	570.67
15.	Mango leaves	17.00	643.20
16.	Sugarcane top (with mustard cake 10% + molasses 10%)	105.00	352.00

Source: Report NATP Project "Evaluation of Locally available---------diets" (2001-02)

Mechanical densification of fodder in fact is a value adding process. It leads to advantages like reduction in transportation cost, storage space and better safety during transportation (Singh *et al.*, 2004). The available options of densification are bailing, pelleting and briquetting. In Gujarat, Maharashtra and Rajasthan sporadically locally designed vertical bailers are used to prepare the grass bales of 114 x 114 x 86 cm^3 size of 115 kg weight. The main draw back of this machine is the low bulk density (Samanta *et al.*, 2004). In case of pelleting though the densification is very high but it is energy intensive due to chopping, grinding requirements of fodder. The third option is briqueting technology by use of pressure and heat, though it densify fodder 10-12 times but it has been observed that none of these technologies can handle efficiently the feed material having large particles and higher proportion of crop residues.

Why complete feed block

Simply by making provision of adequate roughage to meet the exigencies by shifting cereal straw roughages from surplus or growing areas is not enough to overcome the scarcity of fodder in general and impact of drought in specific. In fact we are required to ensure adequate balanced nutrition to the animals to maintain body weight above critical levels including allowance for growing stock or pregnant and lactating animals. Thus, combining of densification with concept of complete feed is a viable proposition for developing fodder banks to arrive at sustainable livestock production in the arid and semi arid areas. A complete feed machine devised by the Indian Agricultural Research Institute (IARI) seems to be appropriate to meet these objective of provision of complete feed provided to the animal in densified form (Singh *et al.*, 2002).

It is not only that complete feed blocks can be transported over any distance conveniently without any loss to provide adequate balanced feed to the animals at the time of scarcity but the processing involved in formation of blocks is also reported to considerably improve the feed consumption, growth rate and efficiency of feed utilization both from concentrate and roughages (Sharma and Garg, 2005; Anonymous, 2005; Kumar *et al.*, 2004; Nagpal and Jabbar, 2005).

Complete feed technology

The "Complete Feed Block" is a densified product compressed at different pressures depending upon species, containing both roughage and concentrate in desired proportion to fulfill the nutrient requirements of target animal production (Samanta *et al.*, 2003). The block may be square, quadrangle or circular depending upon the dye used in the pressure applying machine.

Advantages of complete feed blocks

1. Non-conventional, less palatable feeds can be used which help in reducing the total cost of feed.
2. During storage, prepared feed blocks required one third less space as compared to their loose form
3. Transportation of such blocks is convenient and trouble free.
4. Reduces transportation and storage cost.
5. The palatability, voluntary intake and nutrient utilization of low-grade roughages are increased.
6. In the form of blocks the fermentation of feed is better in the rumen of animal because concentrate stick to roughage on application of high pressure during block formation resulting into stable rumen environment.
7. Prevents loss of lighter nutrients resulted from sneezing.
8. Reduces the wastage of feed material by animal during feeding.
9. Saves labour on feeding management.

Complete Feed Block Making Machine

(A) Structural and operational details of the machine

Complete feed block making machine can compress all kinds of feed material to square shape (20 cmx20 cm) and of desired thickness and weight. The working pressure of the machine can be raised up to 6000 PSI, the output capacity of the machine is 30-40 blocks / hour. The bulk density of the roughage based feed blocks prepared from this machine has been found to be 4 to 5 times more than the original feed. It would increase even more if fixed with binders and concentrates. the machine is powered with 15 H. P. electric motor and it consists of following components.

1. Frame
2. Power pack
3. Hydraulic Cylinder
4. Electric control panel

(B) Advantages of Machine

This machine is very simple in operation and only one person (skilled/unskilled) is required to regulate entire operation of the machine. The working pressure of machine is adjustable (up to 6000

PSI). It can compact all kinds of feed materials to square shape (20 X 20 cm.) and of desired thickness and weight. The blocks retain their shape at compaction for long duration. Guar straw, mustard straw and sugarcane bagasse can form stable blocks when densified at a natural dried moisture content and 418.5 kg/cm^2 pressure. The increase in the moisture content of the materials caused very high resiliency, hence, instability of the compacted blocks.

Fig. 1. Complete feed block machine

(C) Cost of operation of animal feed block making machine

The cost components in feed block formation with the animal feed block making machine can be divided into fixed cost (depreciation, interest on investment, taxes, insurance) and operational cost (repair and maintenance, lubrication, electrical charges, wages of unskilled labour).

Major components for operational costs are given in table 2 based on these the cost of complete feed blocks comes to Rs. 375/MT.

Table 2. Cost component for operation of complete feed block making machine

1.	Price of the machine	Rs. 425000
2.	Capacity of the machine	2 MT per day
3.	Life of the machine	20 years
4.	Interest rate	15%
5.	Working days per year	300
6.	Working hours per day	10
7.	Salvage value	5%
8.	Weight of the machine	3.5 ton
9.	Overall size of the machine	17 ft X 6 ft X 8 ft
10.	Power requirement	15 HP, 3 Phase
11.	Labour requirement	1

Source: Report NATP Project "Evaluation of Locally available---------diets" (2001-02)

(D) Break Even Analysis

The break even analysis shows the profits from the excess of the total revenues over total cost. The break even point identifies the volume of the activity at which total revenues equals total cost. This is often an important point for decision making because it represents a minimum acceptable level of operations. The break even point is defined as the level of operation at which total revenues (TR) equals total cost (TC). It is that volume of output where fixed cost (FC) and variable costs (VC) are covered but no profit exists. The relationship between cost and revenues for different volumes of output indicates that break even point for this machine is 105 tones. This means that the machine will give profits when the volume of production exceeds this figure.

Effect of pressure on bulk density

The study of effect of different pressures viz. 2000, 3000 and 4000 PSI on bulk density considering physical characteristics of blocks prepared for different roughages like grasses, straws, stovers, tree leaves etc. in combination with formulated concentrate mixture (60: 40) showed that with decreasing pressure the expansion of compressed feed material increased with passage of time and consequently bulk density decreased. Bulk density of grass and concentrate mixture (60: 40) increased by 8.41, 7.65 and 7.31 fold at 4000,3000 and 2000 PSI, respectively as compared to loose form. Pressure application appears to be the most critical factor in making of complete feed block (Kundu *et al.*, 1998). Further studies (Samanta *et al.*, 2004) have also shown that blocks prepared from straws like wheat and paddy may get broken even while dropping from 1 meter height, therefore, require a suitable binder for making blocks for complete feeds having such roughages in composition.

Scope of incorporation of unpalatable feed and fodder

In spite of high nutritive value some of the feeds are unpalatable due to presence of some phyto toxins that restricts its use in livestock feeding. But by incorporating such feeds in complete feed blend these can be efficiently utilized in livestock feeding as it prevents selective eating by animals. Certain non conventional feeds such as cotton straw, mustard straw, gram straw, tumba seed cake, mesquite pods, mahua cake etc. have been successfully incorporated in the complete feed blocks at different levels (Samanta, 2004; Anonymous, 2005). Thus, the non-conventional roughages and concentrate having low palatability can be judiciously incorporated in complete feed blocks for sustainable animal production.

Fig. 2. Sheep eating complete feed block

Reduction in transportation cost

The very wide inequality among areas regarding availability of roughages throughout the country in general and during natural calamities in specific require little or more transportation of straws, stovers and grasses from the area of surplus to the affected pockets but as such the transportation of roughages has always been observed to be non viable proposition due to their low bulk density, commonly ranging between 65-75 kg/m³ (Yadav, 1990). The densification through complete feed block machine enhances bulk density by 5-8 times for roughages alone and 4-6 times for complete feeds containing roughages and concentrate in the ration at 60:40 ratio (Samanta, 2004).

Table 3 shows the effect of densification on reduction in the transportation cost for various distances for grass and straws transported in loose and block form. The reduction in transportation cost could bo revealed to the extent of 94% as compared to loose form as a result of enhancement in bulk density. In case of straws the reduction in cost could be observed to the extent of 77% while carrying the material to a distance of 100-400km. Thus, from these observations it appears that densification is suitable answer for transportation of roughages not only in the state of emergency but in routine also by combining it with the concept of complete feed for proper disposal of roughages at production site and ensuring of adequate balanced nutrition to the animals of scarcity prone area.

Table 3. Effect of densification on reduction of transportation cost

Feed materials	Block density (kg/m³)	Capacity (qt/truck)	Transportation cost (rupees / qt.)			Reduction in transportation cost (%)
			100 km	200 km	400 km	
Grass loose	25	7.25	620	896	1379	
Grass block	453	131.37	34	49	76	94
Wheat straw loose	94	27.26	165	238	366	
Wheat straw block	421	122.09	36	53	81	77
Paddy straw loose	89	25.81	174	251	387	
Paddy straw block	395	114.55	39	57	87	77

Source: Samanta (2004)

Reduction of storage space

For developing fodder banking the requirement of space for storage is also a matter of serious concern and as such most of the roughages used in animal feeding require huge space for storage. The reduction in storage space for various roughages as a result of compaction could be well appreciated from the figures shown in Table 4.

The increase in bulk density on making complete feed blocks facilitate convenient and economic handling, storage and transportation (Verma, 1996). Therefore, this technology could be applied as viable component of "Famine Feed Bank" in drought prone areas of arid and semi arid parts of the country.

Table 4. Reduction in storage space

Feed resources	Space required per qt. in loose form (m^3)	Space required per qt. in block form (m^3)	Reduction in space requirement (m^3/qt.)
Grass	3.89	0.22	3.67
Wheat straw	1.06	0.24	0.82
Paddy straw	4.04	0.25	3.79
Black gram straw	1.69	0.18	1.51
Ground nut haulm	1.78	0.17	1.61
Stylo straw	3.54	0.21	3.33
Sorghum stover	2.36	0.18	2.18
L. leucocephala leaves	1.20	0.13	1.07

Source: Samanta (2004)

Table 5: Intake and digestibility of complete feed blocks

Blocks type	DMI kg/100 kg.	DMD%	Animal species
Mustard straw based	3.23-3.78	48.86	Sheep
Bajra kadbi based	3.38-4.23		Sheep
Mustard straw with tree leaves based	2.50-3.73	42.5	Sheep
Natural grass based	2.58-3.08	48.99	Hybrid calves
Natural grass based	2.90-3.06	60.10	Goat
Mesquite pods based	3.32-3.64	64.58	Sheep
Groundnut hull based	3.514-4.06	58.15	Sheep
Groundnut straw + Wheat straw based	3.99-4.09	62.71	Calves

Source: Report NATP Project "Evaluation of Locally available-----diets" (1999-2005)

Acceptability of complete feed block by animals

A variety of locally available straws, grasses and other non conventional feed have been used as major roughage source in complete feed blocks successfully. The values of intake and digestibility recorded in various species for different combinations of complete feed blocks have shown that as such there is no adverse effect on intake and digestibility due to compaction of complete feed and the non conventional low palatability feeds can be efficiently utilized in the feeding of animals for economic animal production. The intake and digestibility of dry matter of some of the combinations of complete feed blocks have been presented in table 5 to reveal the acceptability of complete feed blocks by different species of animals.

Fig. 3. Rathi heifers eating complete feed block

Conclusion

Livestock farming is the major component of arid agriculture as it makes arid rural economy more viable and sustainable. The majority of basal roughages and concentrates required for feeding of livestock originate in the areas where crop husbandry has been intensified. This unequal natural distribution require transportation of both roughage and concentrate from production area to the area of consumption. On account of high bulk density concentrate do not pose problem but shifting of roughage is a non viable proposition due to low bulk density. Thus, densification is dire need to minimize cost of transportation of roughages. Further, for sustainable livestock production, the need of adequate balanced nutrition to animals for targeted production arises. The densified block form of complete feed not only overcomes the problems associated with the transportation, handling and storage of feed but also provide adequate balanced nutrition to the animals at the time of scarcity to sustain productivity. Thus, "Complete Feed Blocks" feed processing technology may be foreseen as viable tool that can boost up the economic gains of livestock owners to change the socioeconomic scenario of arid and semi arid area through improved sustainable livestock production.

REFERENCES

Anonymous (2005). Final report of NATP project entitled "Evaluation of locally available feed and fodder to improve quality and formulate complete economic rations with high roughage diet." submitted to ICAR, New Delhi.

Kumar, P.; Lohan, O.P. and Yadav, K.K. (2004). Effect of crop residues based complete feed blocks on nutrient digestibility and growth in buffalo calves. Indian J. Animal Nutr. 21 (1): 30-33.

Kundu, S.S.; Das, M.M. and Sharma, D.D. (1998). Densification for block making and evaluation of calcium hypo chlorite and alkali treated wheat straw based feed block in comparison to urea treated straw feed. Buffalo J. 3: 291-302.

Nagpal, A.K. and Jabbar, A. (2005). Productivity of lactating camels on complete feed blocks. *Indian J. Animal Nutr.* 22(2): 102-106.

Nagpal, A.K.; Jabbar, A.; Singh, G.P. and Sahani, M.S. (2005). Evaluation of complete feed blocks in draught camels in arid ecosystem. . Indian J. Animal Nutr. 22(1): 61-63.

Samanta, A.K., Singh, K.K., Das, M.M. and Kundu, S.S. (2003). Complete Feed Block a new approach for handling and feeding bulky feed resources. Indian Dairyman.55:57-59.

Samanta, A.K., Singh, K.K., Das, M.M. and Kundu, S.S. (2004). In: Complete Feed Block- A balanced ration for livestock. IGFRI, Jhansi

Singh, A. Panwar, J.S.; Kumar, A.; Jha, S.K. and Pandya, A. (2002). Development of densification machine for crop residues based animal feed blocks. J. Agricultural Engineering,39:9-15.

Singh, Nawab; Chaudhary, V.B. and Sahoo, P.K. (2004). A comparison of two forms of complete feed i.e. mash v/s pelleted with regards to nutrients intake by goats. Proc. of XI Animal Nutrition Conference, held at Jabalpur from 5-7 January, P. 85.

Thole, N.S., Sharma, D.D., De Wit, J. and Singhal, K.K. (1991). Proceedings international workshop on "Feeding of Ruminants on Fibrous Crop Residues" held at NDRI, Karnal from February 4-8, 1991. Pp 344-357

Verma, A.K.; Mehra, U.R.; Dass, R.S. and Singh, A. (1996). Nutrient utilization by Murrah buffaloes (*Bubalus bubalis*) from compressed complete feed block. *Indian Feed Sci and Technol.* 59 : 255-263.

White, D.A. (2000). Annals of Arid Zone, 39 (3): 211-230.

Yadav, K.K.; Rathi, C.S. and Lohan, O.P. (1990). Effect of compaction of roughage based complete feed on digestibility and rumen parameters. *Indian J. Anim. Nutr.* 7 : 27-30.

9

Processing and conservation of feed resources for livestock feeding during drought

Harpal Singh

Central Arid Zone Research Institute, Jodhpur 342 003

Drought and its impact

Over the past few years, there has been a worsening incidence of the natural calamities increasing the vulnerability of the communities across the globe. One of the recurring natural calamities is drought and its impacts on the affected population are visible on slow growth rate and agricultural production. Indications of climatic changes reveal that vulnerability of drought may increase in coming years. Drought is a perennial and recurring feature in many parts of India, particularly arid zone. As per the reports from the Govt. of India about 68% area of the country is prone to drought of varying degrees. The deviations in climatic and rainfall patterns in the country have triggered different effects on people's livelihoods. It can not be seen mere by a physical phenomenon. Each drought leaves a unique set of impacts, depending on its severity, duration, spatial extent, and also on ever-changing socio-economic conditions. The impact is on huge population spread over a large geographical area in a State or over many States for varying duration of time ranging from few months to several years. The impact of drought on livestock is manifested in four ways. (i) mortality (ii) loss in productivity (iii) health of animal and (iv) loss in fertility.

The drought of 2002 -03 arose following an unprecedented failure of rains from south - west monsoon in 2002, particularly in July 2002, when the country as a whole received 19% deficit in rainfall, making that month as the driest over July in the annals of Indian Meteorological records. Arid and semi-arid regions constituting 40% of geographical area are more prone to drought. The frequency of drought in 1/3rd arid areas is twice in five years, where as it is only once in five years in 2/3rd semi-arid region. Drought poses severe scarcity of water, fodder and food grains to begin with followed by nutritional disorders and sometimes casualties in extreme cases. During the drought of 2002-03 sheep and camels followed by cows were the most vulnerable animals (Table 1). They perished in large numbers ranging between 10 to 32% of the population. 25 to 50% of cows were abandoned due to starvation or put in charity centres because of acute shortage of fodder and finances.

Table 1. Growth and composition of livestock in Rajasthan (figures in million)

Year	Cattle	Buffalo	Sheep	Goat	Camel	Others	Total
1951	10.787	3.045	5.387	5.562	0.341	0.399	25.5
	(42.3)	(11.9)	(21.1)	(21.8)	(1.3)	(1.6)	(100)
1961	13.136	4.019	7.361	8.052	0.570	0.371	33.5
	(39.2)	(10.0)	(22.0)	(24.0)	(1.7)	(1.1)	(100)
1972	12.470	4.593	8.556	12.162	0.745	0.353	38.9
	(32.1)	(11.8)	(22.0)	(31.3)	(1.9)	(0.9)	(100)
1977	12.896	5.072	9.938	12.307	0.752	0.394	41.3
	(31.2)	(12.3)	(24.0)	(29.8)	(1.8)	(1.0)	(100)
1983	13.504	6.043	13.431	15.480	0.756	0.436	49.6
	(27.2)	(12.2)	(27.1)	(31.2)	(1.5)	(0.9)	(100)
1988	10.916	6.340	9.913	12.593	0.721	0.418	40.9
	(26.7)	(15.6)	(24.2)	(30.8)	(1.2)	(1.0)	(100)
1992	11.639	7.765	12.497	15.352	0.744	0.454	48.5
	(24.0)	(6.0)	(25.1)	(31.8)	(1.5)	(0.9)	(100)
1997	12.159	9.756	14.312	16.937	0.668	0.490	54.3
	(22.4)	(18.0)	(26.3)	(31.2)	(1.2)	(0.9)	(100)
2003	10.871	10.511	9.759	16.774	0.497	0.480	48.9
% change	(22.2)	(21.5)	(19.9)	(34.3)	(1.0)	(0.10)	(100)
1977-03	-10.6	7.7	-31.81	-1.0	-25.6		-10.0

Goats and other small animals are also affected by drought, but relatively mildly compared to large animals. They are mainly grazing animals and depend on common property resources i.e. grazing lands, wastelands, forests etc. Moving to better-endowed areas saves them. Migration of small animals is traditional practice in Rajasthan. In case of widespread drought, they are taken out of the State, and in the worst situation sold for slaughter, as there is a round the year market for them. All caste, class and economic categories of household keep small animals for milk and ready cash. However, their health is affected by loss of vegetation of their choice. When rain comes, there is a change in type of vegetation for grazing and few animals cannot adjust and die (10% of the total number, as reported by people).

Livestock is the main source of income in arid zone as cropping is not a dependable proposition. However, beside water and food grains, scarcity of fodder is one of the main issues encountered during drought period. The livestock being an important asset and equally affected by drought the state also arranges fodder for cattle incurring the heavy expenditure on transportation and purchase of fodder (Table 2).

Fodder and feed availability

Availability of fodder and feed resources in the region is usually much below the requirement. According to one estimate, the availability of green fodder in the western dry zone of the country is 26.83% of the requirement, that of dry fodder 12.41% and concentrates 5.86%. The situation worsens during the drought, especially in the arid areas. Estimation for western Rajasthan suggests that with current practices the fodder deficit during normal years would be as high as 60% of the demand, and might range from 55% in the western districts of Bikaner, Jaisalmer and Barmer to 69% in the central districts of Jodhpur, Nagaur and Churn, and 72% in Pali, Jhunjhunu and Sikar districts. During drought years the overall deficit might go as high as 76% of the demand, and range from 76% in western districts to 81% in the central districts and 82% in the eastern ones. The northern districts of Ganganagar and Hanumangarh are likely to face a deficit of 24% during normal years and 50% during drought years.

Table 2. Item wise expenditure on drought relief works in Rajasthan (million Rs.)

Item	1999-00	2000-01	2001-02	2002-03	2003-04	2003-04	2004-05
	AE	AE	AE	AE	BE	RE	BE
Gratuitous relief	0.00	27.9	18.45	171.45	29.52	63.0	19.35
Drinking water	8.1	10.8	19.35	17.55	31.95	92.25	22.05
Supplementary nutrition	0.00	0.00	0.00	0.00	0.00	293.4	0.00
Fodder supply	18.45	213.75	162.45	1561.5	504.0	2062.35	122.4
	(0.76)*	(4.3)*	(5.1)*	(36.7)*	(28.7)*	(23.5)*	(5.0)*
Animal health	0.00	0.00	0.00	24.75	49.05	11.25	19.35
				(0.6)*	(2.8)*	(0.1)*	(0.8)*
Public health	0.00	27.9	0.00	203.4	0.00	342.45	0.00
Other expenditure	2648.25	4690.8	2959.65	2274.75	1143.9	5901.3	2277.45
Total	2675.25	4970.7	3159.65	4253.85	1758.15	8766.0	2461.05

Actual Expenditure (AE); Budget Estimate (BE); Revised Estimate (RE) *Figures in parentheses indicate %age of total expenditure.

In almost all the districts of western Rajasthan fodder situation during 2002-03 was very bad. The major sources of fodder during normal years are the crop residues, as well as the natural vegetation in current fallows and uncultivable lands. The permanent grazing lands (Orans) are usually in a severely degraded state, are heavily encroached, and are not always counted for fodder availability. The exceptions are the very few managed ones. Under such a scenario, the almost total crop failure during Kharif 2002, and drying of natural vegetation in the fallow lands meant a fodder crisis in many districts.

Types of fodder available

During normal rainfall year crop residues viz., bajra (pearl millet) *kadbi,* dry leaves & plant of green gram, mothbean, clusterbeans, wheat straw etc. become the main source of animal feed. Besides, grasses, tree based top feed, *pala* (leaf fodder of *zizyphus numularia), loong* (leaf fodder of *Prosopis cineraria)* and wheat straw are also available for consumption. However, during drought period the availability of crop residues is reduced drastically due to crop failure.

Regeneration and re-growth of natural vegetation are also affected severely under drought, especially because of non-availability of moisture and due to higher incidence of browsing, lopping and cutting of trees and shrubs for fodder and fuel purposes. For example, in Bikaner district regeneration of *Cenhrus setigerus (Dhaman), Cenchrus biflorus (Bhurat)* and *Dactyloctinum sindicum (Gathiyan)* grasses was badly affected during drought of 2002-03. *Panicum antidotale (Murath)* failed to sprout properly on the sand dunes, but *Lasirus sindicus (Sewan* grass) performed satisfactorily in sandy habitat. In neighboring Churu district *I. sindicus* failed to regenerate properly, while *C. setigrus* and *C. setigenrus* experienced a reduced vigour and density.

Issues of grazing and fodder availability

Grazing of animals is fast becoming a major societal issue in rural areas of western Rajasthan that often reaches a flash point. With increasing population, fragmentation of crop lands and decreasing fallow there is now fewer areas left for uncontrolled grazing by the large herds of sheep and goats. Most of the permanent pastures *(Orans* and *Gochars)* in the region are in a severely degraded state that makes conditions worse. Pastoralists (nomadic cattle rearers/breeders) who take their animals on annual migration to the greener parts of the state and outside are facing increasing resistance from the farmers which sometimes lead to clashes.

Thus, animals are the most sufferers during drought period both with respect to fodder/feed and water. In situations it becomes imperative to process the available fodder, store and conserve the feed resources for consumption of livestock during drought period.

Indigenous techniques of fodder storage in western Rajasthan

Karai: *Karai* is one of the indigenous methods of fodder storage in western Rajasthan. In this method fodder can be stored safely upto 10 years. The *karai* is made slightly away from the village. Locally available material is used for making the *karai.* It is made in such as way that the stored feed is safe from sun, rain water and high wind storms and also from human and animal's attack. The *karai* is conical in shape with 3-4 m dia at the base and

8 m height (Fig. 1). At the base pearl millet *"bhani"* and mustard straw is kept to safe guard the stored fodder against termite. The *karaiis* covered with *jhumpi* made of pearl millet stalk and is replaced every third year. The *karai* is provided with 0.5 m wide 1-1.5 m, deep trench around it to safeguard against domestic as well as wild animals.

Pachawa: *Pachawa* is one of the indigenous techniques of fodder storage in arid region. The capacity of the *pachawa* is as high as 1000 q. The shape of *pachawa* at the base may be circular, square or rectangular (Fig. 2). The square/rectangular shape pachawa may be consumed slowly with time whereas the circular shape *pachawa* once open is to be chafed in pieces and stored in room/covered space and / or to be sold/consumed. The *pachawa* is erected on a raised floor to safeguard against rainwater and away from the hutments to minimize the risk of fire. While erecting the *pachawa* its longest side should be kept parallel to the commonly blowing winds to avoid any over turning.

Fig. 1. Traditional storage of straw/*pala (Karai)*

Fig. 2. Traditional storage of pearl millet stalk and other crop residues *(Pachawa)*

Muştard and sesamum's stalk alongwith other dry grasses are spread over (1-1.5 m thickness) raised floor to safeguard against termite. Total height of the pachawa is kept around 6-8 m whereas width is kept around 4-5 m. After attaining the height (6-8m) the top surface is made like inverted "V" and is covered with thatched roof (V -shape) made using the pearl millet stalks and other locally available materials. This roof is replaced every third year. The roof cover is extended all around in such a way that the stored fodder do not come in contact of rain water. The *pachawa* is provided with dry branches of thorney plant upto 2-3 m height and an open trench 0.5 m wide and 1-1.5 m deep around it provide safety against domestic as well as wild animals.

Densification and brequetting of fodder/feed

As fodder is a low density material its transportation, handling and storage not only become problematic but also require large storage space. India produces 540 million tones of crop residues annually, which could be used for feeding the country's animal production. There are regions with deficit and excess production of these crop residues in the country. The feed management, with roughage diety would improve manifold if processes are developed such that deficit and disaster prone areas could be stocked with processed animal feed.

Baling/densification of grasses/crop residues

The grasses/crop residues can be densified/baled to 1/3rd of its original volume and stocked/stored for use during lean period/deficit areas(drought prone areas). The baling is generally done slightly at high moisture content. A tractor drawn baling machine was developed at CAZRI (Fig. 3). Block/bale 50 x 50 x 50 cm size can be made using this machine. Feed of processing units are available for polarizations of concentrate. However, these units can not handle a feed mixture having large sized straw; high proportion of chopped fodder and crop residues Engineering Division of IARI has developed a machine for moulding animal feed into compressed blocks using a whole range of crop residues, feed supplements and suitable binders.

A prototype machine for compaction of bio-materials using hydraulic cylinders for application of compaction pressure upto 425 kg/cm^2 was designed and developed (Fig. 4). It is powered with a 25 hp electric motor to run its hydraulic system. Properly sized animal feed material is placed in the hopper of the machine. The quantity of feed is calculated according to the volume [20 cm x 20 cm x L (thickness)] and the density desired of the feed block. The output capacity of the machine is 250 kg/hr. The machine can densify the feed material upto $1/4^{th}$ to $1/5^{th}$ times of the original volume. The volume reduction of individual raw feed material is presented in Table 3. From the table it is clear that the machine is capable of reducing the volume of the feed material in the range of 4 to 42 times. The cost of feed block formation is Rs 400 per tonne.

Fig. 3. Tractor drawn straw / grass bailing machine

Fig. 4. Prototype of animal feed block formation machine

Advantages of complete feed blocks

1. Ease in handling of feed materials and saving of labour in feeding
2. Require less transportation and storage space
3. Less transportation and storage cost
4. Reduction in wastage of crop residues
5. Offer scope for incorporation of hard or large-size feed materials into feed blocks
6. Offer scope for ration manipulation according to the requirement of animals

7. Prevent selective eating by animals
8. Useful to induce animals to consume proper proportion of roughages and concentrates for optimum performance
9. Since feed material becomes soft under compression, animals prefer eating blocks
10. Milk yield and fat contents increase on feeding complete feed blocks.

Table 3. Bulk density and volume reduction of raw feeds and blocks at compaction pressure of 300 kg/cm² and moisture content of 12% (wb).

S. No.	Feed/crop residue	Bulk density of feed (kg/m³)	Bulk density of feed block (kg/m³)	Volume reduction factor
1.	Groundnut straw	81.00	553.63	6.8
2.	*Ardu* leaves	51.50	539.06	10.4
3.	Mustard straw	51.48	562.21	10.9
4.	Gram straw	66.48	471.94	7.1
5.	Wheat straw	51.20	305.22	5.9
6.	Wheat straw with 5 % molasses	57.00	443.50	7.8
7.	Sorghum straw	70.00	437.20	6.2
8.	Paddy straw	70.00	279.00	3.9
9.	Forest grass	66.33	503.00	7.6
10.	Pearl millet	119.60	511.10	4.3
11.	*Cenchrus* grass	67.50	486.84	7.2
12.	*Arharstraw*	55.00	689.06	12.5
13.	Sugarcane baggase	61.00	567.50	9.3
14.	Cluster bean	109.00	596.67	5.5
15.	Mango leaves	17.00	643.20	37.8
16.	*Khejri* leaves	67.50	542.00	8.0
17.	*Pala* leaves	65.50	538.40	8.2
18.	Maize *(chopped)*	90.70	520.80	5.7
19.	Sugarcane top with mustard cake 10% + molasses 10%	105.00	422.00	4.0
20.	*Sewan*	91.60	700.00	7.6

21. Groundnut haulm	77.10	500.00	6.5
22. *Moth chara*	84.70	901.00	10.6
23. Cotton stalk	140.00	785.00	5.6
24. Rice bran	361.00	1197.00	3.3
25. Hay *(Bar seem)* chopped	55.00	1050.00	19.0
26. Korean lawn grass *(Zoysia japonica)*	38.00	597.00	15.7
27. *Doob* grass *(Cynodon dactylon)*	15.00	630.00	42.0
28. Maize cob *(ground)*	498.00	843.00	1.7

Brequetting of animal feed

As availability of green fodder is only 27 % against requirement in the western Rajasthan the animals are ill fed particularly during drought period. To overcome the problem the farmers supplement the feed deficit by multi nutrients feed block / concentrate at least in case of milch animals. Batch type multi nutrients feed block making systems are available. For large scale production of feed block (wheat bran, guar brokens, guar-gum, dolomite, mineral mixture, common salt, urea, molasses, water) continuous type machine is required. Experimental unit of the continuous type animal feed block making machine has been designed and developed at the CAZRI (Fig. 5). About 40 blocks (25 x 15 x 7.5 / 5.0 cm size) per hour can be made with this unit.

Fig. 5. Experimental unit of continuous type feed block making machine

REFERENCES

1. Narain P. 2006. Feed and fodder management in arid region. In Proc. of Short Course on "Feeding of livestock during drought and scarcity". Div. of Animal Sciences and Forage Production, CAZRI, Jodhpur, pp: 1– 8.
2. Singh H; Mishra D; Nahar NM. 2002. Final Report on ICAR Ad-hoc Research Project on Energy Census and Resource Assessment of Arid Zone (Western Rajasthan). Division of Agricultural Engineering and Energy, CAZRI, Jodhpur, pp: 1-201.
3. Narain P; Amalkar. 2005. Drought in western Rajasthan – Impact, Copping Mechanism and Management Strategies, CAZRI, Jodhpur, pp: 1-104.
4. Pathak PS. 2005. Importance of feed resources for livestock production improvement in India. In Proc. (chapter XVIII) of workshop on Fodder Oats, Fodder Technology Packages and Small Farm Income Generation held at Kathmandu, Nepal during March, 8-11, 2005, Editors: D Periyar, J Suttie and S G Reynolds,.
5. Singh H. 2006. Processing and conservation of feed resources for livestock feeding during drought. In Proc. of Short Course on "Feeding of livestock during drought and scarcity". Div. of Animal Sciences & Forage Production, CAZRI, Jodhpur, pp: 105 – 111.
6. Annonymous. 2003. Agricultural Statistics at a glance. Directorate of Economics and Statistics, Govt. of Rajasthan, Jaipur.
7. Annonymous. 2004. Statistical abstract of Rajasthan Directorate of Economics and Statistics, Govt. of Rajasthan, Jaipur.
8. Singh GP; Saini N. 2002. Role of anaerobic fungi in fiber digestion and its special significance of camel nutrition, Indian Dairyman, 54: 64-68.
9. Kavia ZD. 2003. Uttar Paschimi Rajasthan Me Chara Bhandaran Kee Tucknique (in Hindi). Maru Krishi Chayenika (January to December issue), CAZRI, Jodhpur, pp: 57-62.
10. Venkateshwarlu J. 1992. The scenario for Arid Rajasthan. FORAGE 2000 AD. CAZRI,Jodhpur,India
11. Patil SA; Suresh C; Alagunlagi. 2005. Forage production and feeding during scarcity. Livestock Feeding Strategies for Dry Regions. International Book Distribution Company, Charbagh, Lucknow, India.
12. Annonymous. 2006. Gopalan se svavlamban va jaivikkheti, Prakashak, Rajasthan Goseva Sangh Kanhaiya Goshala, Pal Gaon, Pal Road, Jodhpur.

13. Singh A; Jha SK; Panwar JS. 2004. Success Story No. 2 on "Technology for Animal Feed Block Formation Machine" (ICN: 10/2004), IARI, New Delhi-12, pp: 1-14.
14. Punia BS; Lailer PC. 2006. Feeding of buffaloes in arid and semi-arid regions. In Proc. of Short Course on "Feeding of livestock during drought and scarcity". Division of Animal Sciences and Forage Production, CAZRI, Jodhpur, pp: 144 – 151.

10

Weeds as a Source of Fodder in Hot Arid Zone-A Review

J.P. Singh, B.K. Mathur, V.S. Rathore and R.K. Beniwal

Central Arid Zone Research Institute, Regional Research Station, Bikaner-334004

Introduction

Arid grazing lands suffer from two basic deficiencies viz., poor legume component in the natural pasture vegetation and poor soil fertility (Singh and Shankar, 2002), which leads to lower productivity and quality of forage from the grazing lands. Herbaceous species comprising of grasses, legumes and forbs are an integral part of the life support system for grazing animals during normal as well as in the feed scarcity periods. The rangeland vegetation in the Thar Desert is predominantly therophytic and annual weeds constitute the bulk of vegetation particularly in the rainy season (Singh *et al.*, 2006). They have high grazing value and add a variety of diets to livestock, particularly preferred by small ruminants viz., sheep and goats and wildlife. Weeds are able to tolerate long droughts and also provide soil cover (Bhandari, 1990). In arid zone, most of the annual weeds are short lifecycle, rainfed, which emerge with the first showers of monsoon and mature within 60 to 75 days. Monsoon forages perishes by first quarter of November month and their after starts lean period for grazing cattle in arid region, and cattle shows negative growth from December onwards with signs of debility and malnutrition if not provided supplementary feeding (Mathur *et al.*, 1991). The pastoralists also harvest some of the weeds during the post-monsoon and used for feeding the animals in the lean period.

Legume component has an important role in grassland/grazing land production and utilization. Nutritional attributes of weeds of pasture/grazing land indicate that they are important to improve quality of range biomass leading to better health, growth and productivity of animals particularly during lean period. Looking to the role of herbaceous vegetation in the grazing based livestock rearing, there is need to assess their production potential and nutritive value. Moreover, since most of these species are fed to animals after mixing with other feeds, there is also need to workout the feed ratio of these species with other feed ingredients. Information on weeds of fodder importance in arid region is diffused in literature and efforts are made to scan through the available literature in the present review.

Grazing Preference of Weeds by Animals

Many factors influence the quantity and quality of plant material eaten by grazing animals (Arnold, 1964). Few species preferred or rejected by cattle are grazed by sheep goats and camels. In general, weeds are more grazed by goats followed by sheep and cattle. In a study of grass-legume cafeteria of 18 perennial grasses and 18 herbaceous/shrub legumes, goats devoted more time (24%, 29%, 25.6% and 28% grazing time in monsoon, post monsoon, winter and spring, respectively) grazing annual/perennial weeds (Singh and Shankar, 2000) followed by sheep (Singh and Shankar, 2002) and cattle (Singh and Shankar, 1994) (Table 1). It is observed that annual grasses were more preferred by animals in the monsoon over sown perennial grasses. It is believed by the inhabitants that after grazing/feeding some of the species the quantity and quality of milk is improved (Singh and Rathore, 2005).

Table 1. Percent time spent by animals in pasture grasses, legumes and weeds in a grass legume pasture cafeteria

Animal	Species	Season				Reference
		Monsoon	Post monsoon	Winter	Spring	
Cattle	Grass	68.0	59.2	50.6	57.5	Singh and
	Legume	15.0	25.8	41.6	34.1	Shankar,
	Weeds	17	15.0	7.8	8.4	1994
Sheep	Grass	29.3	20.0	47.0	67.1	Singh and
	Legume	54.3	65.0	38.2	16.6	Shankar,
	Weeds	16.4	15.0	14.8	14.3	2002

Goat	Grass	17.4	12.7	23.0	37.7	Singh and
	Legume	58.0	57.9	51.4	33.6	Shankar,
	Weeds	24.6	29.4	25.6	28.7	2000

Preferential grazing/browsing behavior of sheep and goats was studied by introducing 4 sheep and 4 goats of about 6 month age in a silvi-pastoral system for two weeks in September after the first cutting of grass. The mixed pasture contained *Lablab purpureus*, re-sprouted *Lasiurus sindicus* and *Cenchrus ciliaris*, some weed flora as well as woody component consisting *Colophospermum mopane* and *Hardwickia binata*. The animals preferred weed flora, i.e., *Cyprus rotundus, Eleusine indica, Celosia argentea, Eragrostis* sp., *Amaranthus* sp. and *Tribulus terrestris*, followed by *L. sindicus* and *C. ciliaris, L. purpureus* was not preferred (Anonymous, 2005). A list of important weed species of hot arid region eaten by animals is given in Table 2.

Table 2 Important weedy herbaceous plant species in arid region.

Species	Local name	Remarks
(a) Leguminous		
Cassia italica (Mill.) Lamk. ex Anderws	Bhinda-anwal, Goranwal	Ripe pods and seeds eaten by sheep and goats.
Goniogyna hirta (Willd.)Ali	Pilio-bakario	Eaten by all animals
Indigofera argentea Burm.	Seyyan	Grazed by all animals and supposed to increase milk yield.
I. cordifolia Heyne	Bekario, Bekar, Meh-phuli	Preferably grazed by small ruminants and supposed to be a good fodder for livestock.
I. linifolia (L.f.) Retz. var. *linifolia*	Lambio bekrio	Grazed by all animals
I. linnaei Ali	Bekario	Grazed by all animals
I. sessiliflora DC.	Bekaria.	Eaten by all animals and supposed to increase milk yield and fat content in animals.
Rhynchosia capitata (Heyne ex Roth) DC syn. *R. aurea* (Willd.) DC	Batt, Papri	Eaten by sheep, goats and camel.
Tephrosia hamiltonii Drumm.	Bansa	Eaten by camels

T. purpurea (L.) Pers	Biyani, Bisoni	Grazed by sheep and goat
T. villosa (L.) Pers.	Ruvali-biyani	Eaten by sheep and goats
(b) Non-Leguminous		
Arnebia hispidissima DC. (Boraginaceae)	Ramasio, Rambio	Eaten by sheep and goats.
Bergia sufruticosa (Delile) Fenzl. (Elatinaceae)	Rai, Rawan, Kankaro	Grazed particularly by sheep and goats.
Blepharis linariaefolia Pers. syn. *B. sindica* (Acanthaceae)	Untkantalo, Bhangari	Eaten by cattle and goats
Boerhavia diffusa L. (Nyctaginaceae)	Chinawari	Eaten by all animals
B. elegans Choisy	-	Grazed by animals
Borreria articularis (L.f.) Mill. (Rubiaceae)	Agio	Grazed by all animals
Chenopodium album L. (Chenopodiaceae)	Bathua	Used as fodder
Citrullus colocynthis (L.) Schrad. (Cucurbitaceae)	Tumba	Fruits eaten by goats and camel
Convolvulus arvensis L. (Convolvulaceae)	Haranpadi	Eaten by cattle and goats
Corchorus depressus (L.) Christenseon (Tiliaceae)	Bahufali, Cham-ghas	Eaten by goats, sheep and camel and supposed to increase milk yield.
C. tridens L.	Kaglar, Kagaroti,	Eaten by goat, sheep and camel and supposed to increase yield and fat content of milk
Cressa cretica L. (Convolvulaceae)	Rudarvanti	Eaten by camels
Cyperus arenarius Retz. (Cyperaceae)	Dachab	Grazed by all animals
Digera muricata (L.) Mart. (Amaranthaceae)	Lolaru	Eaten by all animals
Euphorbia granulata Forsk. (Euphorbiaceae)	Dudhia, Dudheli.	Grazed by sheep, goat and cattle and supposed to increase milk yield.
Fagonia indica Burm. f. (Zygophyllaceae)	Dhamaso	Eaten by goat, sheep and camel

Farsetia hamiltonii Royle (Brassicaceae)	Hiranchabo	Eaten by sheep and goats.
Gisekia pharnaceoides L. (Molluginaceae)	Balukasag, Sureli	Eaten by goats and camel.
Glinus lotoides L. syn. *Mollugo hirta* Thunb. (Molluginaceae)	Bakda, Hata	Used as sheep and camel fodder
Heliotropium marifolium Koen. ex Retz. var. *marifolium* (Boraginaceae)	Bagphuli	Good fodder for camel, sheep and goat and believed to increase milk yield.
Leucas urticaefolia (Vahl) R.Br. (Lamiaceae)	Panhari	Eaten by sheep and goats
Limeum indicum Stocks ex T. Anderson (Molluginaceae)	Chhapri	Eaten by all animals and supposed to increase milk yield.
Mollugo cerviana (L.) Seringe (Molluginaceae)	Chirighas, Chirikhet	Grazed by animals.
Polygala irregularis Boiss. (Polygalaceae)	Phel	Eaten by animals during scarcity and supposed to decrease milk yield in animals
Portulaca oleracea L. (Portulacaceae)	Kulfa, Lunki kulfa	Eaten by cattle, sheep, goat and camel.
Pulicaria crispa (Forssk.) Benth. & Hook. f. (Asteraceae)	Sonali	Grazed by all animals and supposed to be a good fodder.
Solanum nigrum L. (Solanaceae)	Makoi	Eaten by all animals
Trianthema portulacastrum L. syn. *T. monogyna* L. (Aizoaceae)	Santo, Santhi, Lalsabuni	Used regularly as camel fodder.
T. triquetra Rottl. ex Willd.	Fysur lani	Common in gravelly and saline soils. Regularly eaten by camel.
Tribulus pentandrus Forssk. var. *pentrandrus* syn *T. alatus* Del. (Zygophyllaceae)	Bakda	Grazed by sheep and goats and also stored for lean period feeding to animals.
T. terrestris L.	Chota gokhru, Kanti	Grazed by sheep and goats.
Zaleya govinda (Buch.-Ham ex G. Don) Nair syn. *Trianthema govinda* Buch.-Ham. ex G. Don. (Aizoaceae)	Santhi	Mainly used as camel fodder.

Z. redimita (Melv.) Bhand.	Gulalio-Santo, Ratio gudalio	Grazed by all animals
Zygophyllum simplex L. (Zygophyllaceae)	Lunwo, Luni, Lunk	Used as camel fodder
C Grasses		
Aristida adscensionis L.	Lampro	Grazed by animals in young stage
A. funiculasta Trin. & Rupr.	Lampri	Grazed by animals in young stage
Brachiaria ramosa (L.) Stapf var *ramosa*	Murat	Grazed by animals
Cenchrus biflorus Roxb.	Bhut, Bhurat	Grazed by animals before seeding also used as hay
Chloris barbata Sw.	Jharna	Grazed up to flowering stage
C. virgata Sw.	Chota aranio, Chinki	Reputed to be a good fodder grass of saline areas.
Dactyloctenium scindicum Boiss.	Ganthio	Grazed by all animals and supposed to increase milk yield.
Digitaria ciliaris (Retz.) Koel.syn. *D. marginata*	-	Grazed by animals
Eragrostis tremula (Lam.) Hochst ex Steud	Churalio	Grazed by animals
Ochthochloa compressa (Forsk.) Hilu syn *Eleusine compressa* (Forsk) Asch. & Schweinf	Ganthel, Ghora dhob	Grazed by all animals and ranked as a quality fodder grass
Sporobolus helvolus (Trin.) Th. Dur. & Schinz	Khar ghas	Common in saline area, used as camel fodder
Urochloa panicoides P. Beauv.	Kuri, Kuria	Grazed by all animals

Important Species

Leguminous weeds: In hot arid zone, species of *Indigofera* locally called as *Bekaria* are collected, stored and used as feed to livestock. They are rich source of nutrients and grow on wide range of habitats under low rainfall conditions. *I. cordifolia* is collected and fed to camels throughout the year, either as succulent green or as stored feed (Mathur and Purohit, 1979). Feeding of *I. sessiflora* to sheep is supposed to increase wool production. *Indigofera linnaei* is liked by sheep (Patil, 1966) and grazed by all animals. *Indigofera hirsuta*, which is weed in poor sandy soils, pastures/natural grasslands and wastelands, is grazed by goats and sheep. Its foliage contains 13.7% crude protein (Gupta, 1998). *Goniogyna hirta* is relished by all animals. *Tephrosia purpurea* and *T. villosa* is mainly grazed by sheep and goats. Seeds of *Cassia italica, Indigofera cordifolia, Rhynchosia capitata, Tephrosia purpurea* are also the source of feed to animals.

Non-leguminous and non grassy: The species of *Tribulus, Corchorus, Euphorbia, Heliotropium* etc. are important non-leguminous component of herbaceous natural vegetation of hot arid region. *Tribulus pentandrus* (Plate 1) and *T. terrestris* (Plate 2) are collected and fed to camel, sheep and goats and also to milch cattle during normal and scarcity periods. *Corchorus depressus* is commonly called as *Chamghas* forms source of supply of nutrients to small ruminants and is considered as a famine or scarcity feed. *Corchorus tridens* is collected and fed to goats as stall-feeding. *Euphorbia granulata* grows on open sandy plains and gravelly fields; often covers large patches of sandy ground and preferred by animals. In semi arid region, *Euphorbia hirta* is much relished by cattle and also by sheep and goats and supposed to increase milk yield in cows. *Glinus lotoides* is also collected for stall feeding to sheep (Mertia, 1992).

Boerhavia diffusa is one of the common perennial herbs widely distributed throughout India up to 2500m in the warm valleys of the Himalaya. In its natural condition, it flourishes well in monsoon in wastelands and open grasslands/pastures and grazed by sheep and goats and also by cattle. Traditionally it is used as fodder and seems to be promising non-leguminous forage. In Jamaica, its leaves are given to hogs hence it is commonly called "hogweed". The annual species like *Celosia argentea* occurs throughout India in crop fields, grasslands/pastures and grazed by goats, sheep and also by cattle before seed setting during monsoon and post monsoon season. *Digera muricata* occurs in crop fields and pastures is relished by all grazing animals. It shows considerable value as an annual green fodder. The seeds of some of the weeds like *Tribulus* spp. can also serve as feeds for sheep and goats.

Grasses: The species of *Aristida, Brachiaria, Cenchrus, Dactyloctenium, Ochthochloa* are important grasses from grazing and nutritional point of view in hot arid region. The perennial grass

Eragrostis japonica (*E. diarrhacha*) occurs in wet places and pasture also used as fodder. Mertia (1992) mentioned that a large pasture of this grass was established at village Dana in Jaisalmer particularly for horses.

Plate 1. *Tribulus pentandrus*

Plate 2. *Tribulus terrestris*

Production Potential of Weeds

In both, monsoon and winter, weeds infest sown pastures, crop fields and natural grasslands. Species diversity of monsoon weeds is much higher than of winter weeds. Generally, grasses and sedges dominate monsoon weed flora and dicots winter weed flora. Despite several negative effects of weeds in crop and grazing lands weeds serve as a good source of forage to grazing animals. In arid region production potential of weeds contribute differently to the total biomass of the grazing lands representing different range conditions (Kumar, 2005). Monsoon herbage if grown properly, can contribute up to 500-600 kg/ha dry matters in natural condition. Production of weeds from crop fields is also important as weed biomass often far exceeded their companion crop (Shankar and Kumar, 1984) i.e. biomass was 2 to 4 times more than crop field. These are mostly palatable weeds giving 655 to 1564 kg/ha of dry matter depending upon preponderance of weeds. In leguminous weeds, *Indigofera cordifolia* is important from production point of view. In a protected and unprotected village grazing land in Bikaner *I. cordifolia* contributed 268 kg/ha (43.88%) and 21.33 (6.42%) of herbage yield respectively (Kumar, 1991).

Nutritive Value

Weeds of forage significance contain sufficient amount of nutrients to be considered as livestock feeds. Several workers reported the nutritive value of wild species of forage importance of arid region (Mathur and Karwasra, 1967; Nath and Malik, 1970; Purohit and Mathur, 1970; Mathur *et al.*, 1977; Mathur and Purohit, 1979; Nag *et al.*, 1979; Harsh *et al.*, 1981; Grover and Nag, 1984; Sethia *et al.*, 1987; Singh *et al.*, 1987; Kapoor *et al.*, 1988; Mathur *et al.*, 1988; Mathur *et al.*, 1994, Singh *et al.*, 2006). Chemical composition of various weeds is given in Table 3. Chemical composition, however, is reported to vary in plant parts, age of tissues, inter and intra species variation, age, part of plant, season and habitats. Chemical composition alone is an inadequate indicator of nutritive value, therefore, anti-nutritional factors and nutrient degrability is also important from utilization point of view.

***Crude Protein* (CP):** In general leguminous weeds showed high crude protein content ranged from 6.25 to 14.52 with mean value of 11.56% with exception of 17.32 in *Tephrosia purpurea*. *Indigofera cordifolia*, a common leguminous weed showed higher leguminous weeds also show high range of CP from 3.13 to 19.11%. The common weed *Tribulus terrestris* shows much variation in CP content reported by different workers. Nath and Malik (1970) reported high CP (12.06%) of the young dried plants of *T. terrestris* in pre-flowering stage and found the digestibility coefficient as high as 75.20 of the harvested plants given as sole feed to sheep. In *Fagonia indica* high CP reported in summer (9.41%) and decrease in winter (6.73%) season. *Euphorbia granulata* is also important for nutrition point of view. Badri and Hamed (2000) reported 11.40% CP in it from extreme arid environment in Egypt and also reported large amount of steroid and terpenoid and average amount of flavanoid.

Table 3. Chemical composition (%) of arid zone weeds

	Ash	CP	EE	TCHO	CF	NFE	NDF	ADF	Ref.
1	2	3	4	5	6	7	8	9	10
Leguminous									
Indigofera argentea	9.40	10.80	1.20	78.70	41.30	37.40	70.60	56.20	Singh et al., 2006
I. cordifolia	17.30	9.90	1.40	71.40	33.70	37.70	59.70	49.70	Singh et al., 2006
	19.38	6.25	1.76	-	28.33	44.38	-	-	Mathur and Purohit, 1979
	-	14.52	-	-	-	-	31.90	22.90	Bhatt et al., 2002
I. sessiliflora	13.80	9.50	1.00	75.70	32.70	43.00	56.80	44.90	Singh et al., 2006
Rhynchosia capitata	10.00	12.6	2.60	74.90	18.60	56.30	53.90	31.20	Singh et al., 2006
Tephrosia purpurea	-	17.32	-	-	-	-	54.20	39.10	Bhatta et al., 2002
Average	13.98	11.56	1.59	75.18	30.93	43.76	54.52	40.67	
Non-Leguminous									
Bergia sufruticosa	11.60	8.20	1.60	78.6	32.90	45.70	52.60	35.90	Singh et al., 2006
	10.50	5.31	6.57	-	37.75	39.70	-	-	Singh et al., 1987
Blepharis linariaefolia	19.40	10.00	2.50	-	10.50	57.41	-	-	Mathur et al., 1988

Citrullus colocynthis									
Fruits	12.58	17.81	12.22	-	7.21	50.18	-	-	Harsh et al., 1981
Seed cake	11.20	16.9	3.95	67.95	42.0	25.95	-	-	Mathur et al., 2000
Corchorus depressus	18.90	7.56	2.33	-	32.13	44.04	-	-	Mathur and Karwasra, 1967
C. tridens	12.60	7.40	2.20	77.90	33.90	44.00	65.80	47.70	Singh et al., 2006
Euphorbia granulata	13.70	7.80	3.30	75.20	20.00	55.20	39.90	30.60	Singh et al., 2006
Fagonia indica	13.35	10.31	3.94	57.60	14.79	-	-	-	Harsh et al., 1981
Monsoon	6.71	9.41	2.20	-	41.22	40.46	-	-	Shukla et al., 1987
Winter	5.62	7.50	2.73	-	47.53	36.62	-	-	Shukla et al., 1987
Summer	6.17	6.73	4.54	-	44.60	37.96	-	-	Shukla et al., 1987
Glinus lotoides	13.00	12.80	0.80	-	10.10	63.21	-	-	Mathur et al., 1988
Heliotropium marifolium	17.40	11.40	1.90	69.30	27.70	41.60	50.00	34.80	Singh et al., 2006
Limeum indicum	34.40	6.90	0.90	57.80	22.60	35.20	64.80	49.30	Singh et al., 2006
Polygala irregularis	7.40	10.60	1.00	81.00	48.40	32.70	69.60	57.40	Singh et al., 2006
Pulicaria crispa	11.90	7.70	3.00	77.50	40.70	36.80	53.50	50.80	Singh et al., 2006
Trianthema portulacastrum	-	11.37	-	-	-	-	58.60	49.00	Bhatta et al., 2002
Tribulus pentandrus	14.60	9.90	1.70	73.80	35.40	38.40	58.70	48.30	Singh et al., 2006
Stem	8.45	4.69	5.95	-	50.60	30.32	-	-	Mathur et al., 1977

Leaf	19.60	7.82	2.65	-	29.05	40.89	-	-	Mathur et al., 1977
Seed	15.30	8.44	6.30	-	27.60	42.37	-	-	Mathur et al., 1977
T. terrestris	16.72	12.06	2.61	68.61	27.78	40.83	-	-	Nath and Malik (1970)
		13.13					44.40	33.76	Bhatta et al., 2002
		19.11		62.1	16.75				Singh and Saini, 2002
Stem	3.80	3.13	9.95		45.70	33.32			Mathur et al., 1977
Leaf	21.00	6.25	10.45		22.60	39.60			Mathur et al., 1977
Seed	9.35	5.63	11.95		34.25	38.82			Mathur et al., 1977
Zaleya redimita	17.70	11.20	1.00		10.50	59.42			Mathur et al., 1988
Zygophyllum simplex	18.80	7.07	7.79		38.23	28.10			Mathur et al., 1994
Average	13.87	9.21	4.31	70.86	30.02	42.11	55.79	43.76	
Grassy									
Aristida adscensionis syn. *A. depressa*	-	2.10	-	84.2	42.25	-	-	-	Singh and Saini, 2002
Dactyloctenium aegyptium	-	12.13	-	-	-	-	41.87	29.43	Bhatta et al., 2002
	-	8.32	-	75.7	35.19	-	-	-	Singh and Saini 2002
D. sindicum	15.6	11.8	1.6	71.1	33.4	37.7	78.2	49.6	
Ochthochloa compressa	17.3	10.3	0.9	71.4	38.4	33.0	3.5	56.8	
Average	16.45	8.93	1.25	75.60	37.31	35.35	41.19	45.28	

Mineral Nutrients in Weeds

Macro nutrients

In general, information on macro nutrients in herbaceous non-grassy weeds is lacking except Calcium (Ca) and Phosphorus (P) content in few herbaceous species (Table 4). However, good information on macronutrients contents in important forage grasses was reported by Dhir *et al.* 1984 (Table 5).

Calcium: Generally, the forage weeds are good source of Calcium ranged from 1.00 to 9.60%. *Tribulus terestris, T pentrandrus* and fruits of *Citrullus colocynthis* contain high calcium content. In spite of having a low grazing value for domestic livestock, the fruits are eaten by sheep, goats and camel. Badri and Hamed (2000) also reported much greater Calcium level (3.7%) in *C. colocynthis* from Egypt. Mathur (1996) reported feeding *C. colocynthis* seed cake to sheep and prepared cheaper concentrate mixture for arid livestock from locally available resources. Mathur (1996) further reported high nutritive value of Tumba (*C. colocynthis*) seed cake, particularly protein and minerals viz. calcium and phosphorous.

Phosphorus: phosphorus is important in the dietary mineral elements for livestock, as deficiency of P is common in animals grazing on soils poor in P content. Concentration of P content in weeds ranged from 0.038 to 0.828%. The P content in plant tissues declines with increasing maturity and the rate and extent of decline varies with species (Norton, 1981). In grasses it ranged from 0.127 to 0.224%.

Magnesium: The grasses such as *Brachiaria ramosa, Digitaria ciliaris* (syn. *D. marginata*) showed exceptionally high value of Magnesium content.

Table 4: Macro elements in herbaceous weeds

Species	Ca (%)	P (%)	Reference
Indigofera cordifloia	1.98	0.180	Mathur and Purohit, 1979
Bergia sufruticosa	8.61	0.202	Singh *et al.*, 1987
Blepharis linariaefolia Pers.	5.50	0.390	Mathur *et al.*, 1988
Citrullus colocynthis			
Fruits	3.90	0.828	Harsh *et al.*, 1981
Seed cake	0.29	0.89	Mathur, B.K. 1996
Corchorus depressus	2.37	0.282	Mathur and Karwasra, 1967
Fagonia indica Burm. f.	4.90	0.038	Harsh *et al.*, 1981
Monsoon	1.31	0.120	Shukla *et al.*, 1987
Winter	1.13	0.110	Shukla *et al.*, 1987

Summer	1.30	0.150	Shukla *et al.*, 1987
Glinus lotoides	1.00	0.420	Mathur *et al.*, 1988
Heliotropium marifolium			Sethia *et al.*, 1987
Stem	4.28	0.105	
Leaf	6.04	0.125	Sethia *et al.*, 1987
T. pentandrus syn. *Tribulus alatus*	3.00	0.058	Mathur *et al.*, 1977
Stem			
Leaf	6.80	0.134	Mathur *et al.*, 1977
Tribulus terrestris (Post flowering stage)	4.21	0.245	Nath and Malik. 1970)
Stem	3.80	0.113	Mathur *et al.*, 1977
Leaf	9.60	0.289	Mathur *et al.*, 1977
Seed	5.60	0.519	Mathur *et al.*, 1977
Zaleya redimita	3.20	0.330	Mathur *et al.*, 1988
Zygophyllum simplex	1.23	0.462	Mathur *et al.*, 1994
Average	3.99	0.255	

Table 5. Macro elements in annual grasses (Dhir *et al.*, 1984)

	Ca (%)	Mg (mg/100gm)	K (%)	Na (mg/100gm)	P (mg/100gm)
Aristida funiculata	0.50	0.24	0.35	0.020	0.143
Brachiaria ramosa	0.88	0.64	1.37	0.049	0.224
Cenchrus biflorus	0.72	0.42	1.23	0.021	0.148
Dactyloctenium scindicum	0.86	0.32	0.41	0.149	0.146
Digitaria ciliaris (*syn. D. marginata*	0.72	0.84	1.11	0.035	0.163
Ochthochloa compressa syn Eleusine compressa	0.71	0.26	0.35	0.228	0.127
Eragrostis tremula	0.61	0.21	0.32	0.018	0.203
Average	**0.71**	**0.42**	**0.73**	**0.074**	**0.165**

Microelements

In general, arid zone feeds are deficient in micronutrients like copper, zinc, cobalt and manganese. Information on microelements is lacking in herbaceous weeds except grasses. Sharma *et al.* (1984) provided information on micronutrient contents in important grasses of arid zone (Table 6). The species viz., *Brachiaria ramosa* showed high value of iron and manganese, whereas, *Ochthochloa compressa* (syn. *Eleusine compressa*) has high copper and manganese content among arid grasses (Sharma *et al.*, 1984). There is need to work out the micro nutrients in forage weeds.

Table 6. Micro elements (ppm) in annual grasses (Sharma *et al.*, 1984)

Plant Species	Fe	Mn	Zn	Cu
Aristida funiculata	176	55.4	62.2	42.8
Brachiaria ramosa	444	92.8	47.8	35.3
Cenchrus biflorus	230	67.6	43.8	36.8
Dactyloctenium scindicum	238	73.1	51.1	29.8
Digitaria ciliaris syn. *D. marginata*	216	71.8	51.3	30.7
Ochthochloa compressa syn *Eleusine compressa*	216	89.2	66.2	52.2
Eragrostis tremula	191	87.3	77.3	40.3
Average	244.43	76.74	57.10	38.27

Role of Weeds in Animal Health Care

Many of the weedy species have medicinal properties and supposed good for animal health. The arid zone species viz., *Arnebia hispidissima, Bergia sufruticosa, Blepharis linariaefolia, Boerhavia diffusa, Cassia italica, Convolvulus arvensis, Corchorus depressus, Cressa cretica, Fagonia indica, Farsetia hamiltonii, Gisekia pharnacioides, Glinus lotoides, Mollugo cerviana, Portulaca oleracea, Tribulus terristris, Zygophyllum simplex* etc. are having medicinal value. These species are utilized by the local inhabitants to cure many diseases and ailments as well as in the indigenous system of medicine. In Bengal, the plants of *Boerhavia diffusa* are given to cattle as a medicinal food and are also supposed to increase the quality and quantity of milk. It is fed to horses to cure stomach disorders. *Tribulus terrestris*, locally known as Gokhru or Kanti is commonly used to curer many diseases and ailments. However, Tribulus is suspected to toxicity (Chopra *et al.*, 1965), but its toxicity not reported by local farmers in the region. The species of Convolvulus, which are common in grazing lands, cultivated fields are important for animal health care. The pods and seeds of Cassia italica are much eaten by sheep and goats, which grows wild in the sandy wastelands in arid region. Its seeds have purgative and anthelmintic properties. *Corchorus depressus*, the common weed in open gravelly places has known to have traditional medicinal value as tonic and cooling medicine. Mathur and Karwasara (1967) reported that goats build up condition when they consume *C. depressus*, which is considered as a nutritious feed for them. It also fed to camels to cure stomachache (Takhar, 2004). *Corchorus tridens* fed to camels, which is supposed to increase the hump size. *Cressa cretica* which grows in the saline lands is used in numerous indigenous medicine as anthilmintic, stomachic/tonic etc. *Fagonia indica*, a common weed of sandy to gravelly soils is used in both Ayurveda and Yunani system of medicine as a bitter and astringent tonic. *Farsetia hamiltonii*, a weed of cultivated fields is used as cooling medicine. *Gisekia pharnacioides*, common in wet habitat is reputed for having strong anthelmintic properties. *Mollugo cerev-iana*, a common weed in sandy and fine gravel soils and cultivated fields is traditionally used in fever in arid region. Species of Cleome are the common weed in monsoon season and have medicinal value. Seeds of *Cleome gynandra* are given to horses for stomach disorders. *Citrullus colocynthis* is also important for purgative property of its fruits. Its fruits/seeds are fed to animals by local inhabitants to cure constipation. *Blepharis linariaefolia* syn. *B. sindica* having 12.1% CP in fruits (Mathur *et al.*, 1988) is well known for its tonic property.

Besides, some ill effects on health of goats leading to digestive disorders were also noticed by Mathur (1992) on drought stressed goats fed on monsoon forages. The toxic materials known were alkaloid / punarnavine and potassium nitrate in *Boerhavia diffusa;* rotenoid in *Tephrosia purpurea*; oxalic acid in *Protulaca olearacea, Amaranthus spinosus, Cenchrus setigerus, Borreria hispida, Brachiaria ramosa;* and nitrate in *C. setigerus, B. hispida, B. ramosa;* laxative effect in *Digera muricata*; and photosensitization and liver damage by *Tribulus terrestris*.

Degraded lands for Forage Production

Kumar (2005) rightly mentioned that though perennials are preferred for rehabilitation of degraded lands, but once the seed bank by annuals is added to soil, play a key role in accelerating process of rehabilitation. They provide organic matter to soil facilitating growth of perennials. In this context, leguminous weeds have an important role in degraded grazing lands as they fix high amount of symbiotic nitrogen. In hot arid zone, particularly the species of *Indigofera, Crotalaria, Tephrosia* etc are important for greening the bare areas. They show high genotypic /ecotypic variation and are adapted to specific habitats and niches. *Indigofera argentea* is a typical dune pioneer legume having taproot confined to sand dune. *Indigofera sessiliflora*, a caespitose herb occurs in dry sandy soils particularly in hyper arid region. *Indigofera cordifolia* occurs on the sandy to gravelly ground and cultivated fields and often forms pure patches. These species set lot of seeds and return to the soil seed bank. The species of *Tephrosia* show a rich diversity in hot arid region. *Citrullus colocynthis* grows wild in the area is valuable species for increasing plant cover on sand dune. In saline lands, the adapted halophytic herbaceous species of forage importance viz., *Zygophyllum, Sporobolus, Chloris* etc. have an important and yet unrealized role in increasing the productivity of these lands.

CONCLUSION

Herbaceous weeds of forage importance are important to optimize livestock production as they have potential ability to grow under sub-normal rainfall. Among leguminous weeds, species of *Indigofera* are important for hot arid region. Some of the species like *I. cordifolia* show high seed yield on poor soils. It sets plenty of seeds on maturity and return to soil (Natrajan *et al.*, 2002) and the same in *I. seesiliflora*. The major part of the scattered seeds remain in the soil seed bank and exhibit dormancy which is the natural mechanism for their persistence and ecological maintenance. In non-leguminous weeds the species of *Corchorus*, *Euphorbia* and *Tribulus* are important for high seed yield and soil seed bank in the grazing lands. Among grasses the two arid zone species viz., *Dactyloctenium sindicum* and *Octhochloa compressa* are important for nutritional point of view to small ruminants. One vital aspect of weeds utilization is their medicinal value for animal health, which requires attention to the research workers as medicinal forage. There is urgent need to collection and evaluation of potential weeds from range of habitats for adapted ecotypes. Some of these species possesses agronomic potential, thus there is need to work out the agronomic potential of promising weeds which could be used as multicut fodder crop in sole or as intercrop during less rainfall and drought. Estimation of chemical composition especially macro and micro nutrients, digestibility of nutrients and nutritional value is imperative for better utilization and livestock production point of view.

REFERENCES

Anonymous 2005. Annual Report (2005) Central Arid Zone Research Institute, Jodhpur.

Arnold, G.W. 1964. Factors within plant associates affecting the behaviour and performance of grazing animals. pp. 133- 154. In: J. Crisp (Ed.) *Grazing in Terrestrial and Marine Environments*. Blackwell Scientific Publication, Oxford.

Badri, M. A. and A. I. Hamed. 1999. Nutritive value of plants in an extreme arid environment (Wadi Allaqi Biosphere Reserve, Egypt). *Journal of Arid Environments*. 44: 347-356.

Bhandari, M.M. 1990. *Flora of Indian Desert*. (Revised Edition). MPS Repros, Jodhpur.

Bhatt, R, A.K. Shinde, S.K. Sankhyan and D.L. Verma 2002. Nutrition of range goats in shrubland of Western India. *Asian Australasian Journal of Animal Sciences* 15: 1719-1724.

Chopra, R.N., R.L. Badhwar and S. Ghosh. 1965. *Poisonous plants of India*.Vol. 1. ICAR, New Delhi (2nd Revised and Enlarged edition).

Dhir, R.P., B.K. Sharma and B.K. Datta .1984. Mineral nutrient elements in natural vegetation of arid Rajasthan: I Macro elements. *Annals of Arid Zone* 23: 111-117.

Grover, Sharda and T.N. Nag. 1984. Nutritive value of dried Zygophyll-aceous plants growing in Rajasthan: Chemical composition. *Comp. Physiol. Ecol*. 9: 250-252.

Gupta, R.K. 1988. Non-Conventional Plants for Livestock Feeding. pp. 445-450. In: 3rd *International Rangeland Congress*. Abstract Vol. II. Range Management Society of India, Indian Grassland and Fodder Research Institute, Jhansi, India.

Harsh, M.L., G.R. Purohit, C.S. Mathur and T.N. Nag 1981. Nutritive value of dried terrestrial plants growing in Rajasthan: Chemical composition. *Comp. Physiol. Ecol*. 6 (1): 30-32.

Kapoor, B.B.S., S.K. Mathur and T.N. Nag. 1988. Nutritive status of *Aerva tomentosa* growing in arid zone of Rajasthan. *Oikoassay* 5: 1-2.

Kumar, S. 1991. Assessment of grazing resources and their demand availability status through rain use efficiency analysis in arid land villages. *Forage Research* 17: 45-54.

Kumar, S. 2005. Monsoon herbage-their ecology and potential. pp. 1-15. In: Compendium -National Seminar on *Conservation, Processing and Utilization of Monsoon Herbage for Augmenting Animal Production.* CSWRI, Arid Campus, Bikaner.

Mathur, B.K 1992. Effect of monsoon forage on drought- stressed goats managed on natural hot arid rangeland. In Proceedings *"Fifth International Conference on Goats,* New Delhi, India, 2-8 th March 1992", organized by International Goat Association and ISGP. P 178.

Mathur, B.K 1996. Effect of replacement of cotton (*Gossypium* spp.) seed cake from Tumba (*Citrullus colocynthus*) seed cake and sun dried poultry droppings on the performance of sheep. Ph.D. Thesis, Rajasthan Agriculture University, Bikaner, Rajasthan.

Mathur. B. K. 2003. Livestock: human need for sustainability in arid environment. In: Pratap Narain, S. Kathju, Amal Kar, M. P. Singh and Praveen Kumar (Eds.) *Human Impact on Desert Environment.* Arid Zone Research Association of India and Scientific Publishers (India), Jodhpur, India. pp. 506-514.

Mathur, B.K., Bohra, H.C., Mathur, A.C. and Kaushish, S.K. 1999. Post monsoonal feed and fodder sources in arid region. In "Proceedings IX Animal Nutrition Conference, 2-4th December 1999". Acharya N.G. Ranga Agricultural University, Rajendranagar, Hyderabad. P.101.

Mathur, B.K. A.K. Patel, and S.K. Kaushish 2000. Utilization of tumba (*Citrullus colocynthis*) seed cake of desert for goat production. *Indian Journal of Animal Science* 70 (4): 431-433.

Mathur, C.S. and G.R. Purohit. 1979. Nutritive value of bakaria (*Indigofera cordifolia*). *Annals of Arid Zone* 18: 267-271.

Mathur, C.S. and R.S. Karwasra. 1967. Some nutritive aspects of Cham-ghas (*Corchorus anticorchorus* Recusch). *Ind. Vet. J.* 44 : 525-527.

Mathur, C.S., T.N. Nag and S.C. Goyal. 1977. Nutritive status of Gokhru as livestock feed in the arid zone of Rajasthan. *Food Farming and Agriculture* 9 (1): 11-13.

Mathur, Kavita; S.S. Shekhawat, M. Sethia, V. Singh and T.N. Nag. 1994. Nutritive value of *Zygophyllum simplex* Linn. *in vivo* and *in vitro. Oikossay* 11: 17-18.

Mathur, S.K., B.B.S. Kapoor and T.N. Nag. 1988. Proximate composition of some arid zone plants. *India J. Anil. Nutr.* 5:170-172.

Mertia, R. S. 1992. Studies on Improvement and Utilization of Rangelands of Jaisalmer Region. CAZRI. Jodhpur.

Nag, T.N.; G.S. Mathur and S.C. Goyal 1979. Phytochemical studies of *T. alatus, T. terrestris* and *Agave wightii*: Contents of primary and secondary products. *Comp. Physiol. Ecol.* 4(3) : 157-160.

Nath, K. and N. S. Malik. 1970. Chemical composition and nutritive value of *Tribulus terrestris* L. *Indian Journal of Animal Sciences* 40:434-437.

Natrajan, S., J. Mitra and R.K. Jain 2002. Species description and soil seed bank of *Indigofera cordifolia*. In Abstract. National Symposium on Grassland and Fodder Research in the New Millennium. IGFRI, Jhansi. pp 35-37.

Norton, B. W. 1981. Differences between species in forage quality pp. 89-110. In: Hacker, J.B. (Ed.), *Nutritional Limits in Animal Production from Pastures*. Proceedings of International Symposium held at St. Lucida, Queensland, Australia.

Patil, B.D.1966. Bekaria: Sheep grass of eastern Rajasthan. *Indian Farming* 16:37-41.

Patil, N.V, B.K. Mathur, H.C. Bohra and A.K. Patel 2005.Relative preference index for arid forages for goat and sheep. National Seminar on *"Conservation, Processing and Utilization of Monsoon Herbage for Augmenting Animal Production"*, December 17-18,2005. Organized by Indian Society for Sheep and Goat Production and Utilization, at RRS, CSWRI, Bikaner.pp.190.

Purohit, G.R. and C.S. Mathur. 1970. Nutritive value of Murat (*Panicum turgidum*) grass. *Annals of Arid Zone* 9: 261-267.

Sethia, M, K. Mathur, V. Singh and T.N. Nag. 1987. Nutritive status of *Heliotropium* spp. growing in arid zone of Rajasthan: Chemical composition. *Oikoassay* 4: 25-27.

Sharma, B.K., R.P. Dhir and B.K. Dutta 1984. Mineral nutrient elements in natural vegetation of arid Rajasthan: II Micro elements. *Annals of Arid Zone* 23: 235-241.

Shukla, P.C., H.B. Desai, M.C. Desai, L.P. Purohit, J.S. Patel and R.T. Dangarwala. 1987. Sheep nutrition survey in Banaskantha district. *GAU Res. J.* 13 : 31-40.

Singh, G.P and N. Saini, 2002. Role of anaerobic fungi in fibre digestion and its special significance of camel nutrition. Indian Dairyman 54: 64-68.

Singh, J. P. and Vinod Shankar. 2000. Forage choice of goat grazing in a protein bank cafeteria. *Range Mgmt. & Agroforestry* 21: 1-9.

Singh, J. P. and Vinod Shankar. 2002. Diversity of forms and niches in tropical pasture legumes. In: R.R. Rao (ed.) *Advances in Legume Research in India*. Bishan Singh Mahendra Pal Singh, Dehradun. pp. 47-57.

Singh, J.P. , B.C. Mondal, M.L. Soni and R.K. Beniwal 2006. Annual weeds add to the nutritional level of forage from desert rangelands. *Annals of Arid Zone* 45 (2): 189-193.

Singh, J.P. and V. Shankar 1994. Preferences of cattle grazing in a grass-legume cafeteria including forage shrubs and trees. In: P. Singh, P.S. Pathak and M.M. Roy (eds.) *Agroforestry systems for Degraded lands* Vol.II. Oxford & IBH Pub. Co. New Delhi. pp. 658-665.

Singh, J.P. and V.S. Rathore 2005. Potential Herbage Resources and their availability in arid region. In: Compendium -*National Seminar on Conservation, Processing and Utilization of Monsoon Herbage for Augmenting Animal Production.* CSWRI, Arid Camas, Bikaner. pp 16-25.

Singh, J.P. and Vinod Shankar. 2002. Forage choice of sheep grazing in a protein bank cafeteria. *Annals of Arid Zone* 41: 135-141.

Singh, J.P., B.C. Mondal, M.L. Soni and R.K. Beniwal 2006. Annual weeds add to the nutritional level of forage from desert rangelands. *Annals of Arid Zone* 45 (2): 189-193.

Singh, V; K. Mathur; M. Sethia and T.N. Nag 1987. Nutritional aspects of dried terrestrial plants growing in arid zone of Rajasthan: Chemical composition. *Geobios* 14: 274-276.

Takhar, H.K. 2004. Folk herbal veterinary medicines of southern Rajasthan. *Indian Journal of Traditional Knowledge* 3 : 407-418.

11

Role of Remote Sensing in Mapping Grazing Resources

J.R. Sharma and S. Pathak

Regional Remote Sensing Service Centre, ISRO,
Dept. of Space, Jodhpur - 342 003

Remote sensing due to its synoptic and repetitive coverage helps in cost effective mapping of earth resources. Projects have been carried at national and regional level for mapping of wastelands in general and grasslands/grazing lands in particular. In last two decades number of sensors have been developed and deployed on satellites considering vegetation as a resource of prime importance. In Rajasthan, studies have been carried out for wasteland mapping and also on CPRs.

The repetitive nature of satellite coverage helps in mapping grazing lands in different season thereby not only mapping the spatial distribution and extent but also their seasonal variations. In the context of semi-arid and arid region the season variation is very critical. This paper presents one such study which is carried out for studying dynamics of vegetation in Rajasthan which is largely dominated by grazing/grasslands in western Rajasthan. Studies are also carried out mapping of CPRs and would be presented.

Vegetation is a part of complex eco-system, which includes the plant cover, landforms and features of its surface deposits. In desert, a close relationship exists in a dynamic equilibrium among habitat, vegetation and its environment. Minor changes in the physical environment entail dynamic changes in vegetation. Vegetation differs in sequence of species and varies in sequence of time as well. The dynamic rhythm of vegetation varies in magnitude from extremely arid region of rainfall (100 to 150 mm) of western part of Jaisalmer to eastern part of Rajasthan (200 to 750 mm) and beyond. Blatter and Hallberg (1918-21) viewed that the vegetation types of western Rajasthan have been exclusively controlled by edaphic factors. Biswas & Rolla (1953) and Bharadwaj (1961) also classified desert vegetation taking edaphic factors into consideration.

Most of the area falls under arid climate and is characterized by sandy plains more or less barren of vegetation except in rainy season when multitudes of ephemerals come up and transforms the barren into green carpet. Ephemerals complete their life cycle before advent of summer heat and bulk of the area is once again transformed into open sandy plain, desolate and barren. In general vegetation in arid region is sparse. The bulk of vegetation consists of stunted, thorny shrubs and perennial herbs. In semi-arid region the vegetation mainly consists of grasses, interspersed with few desert shrubs. Depending upon rainwater and phenological events vegetation can be distinctly divided into EPHEMERAL, SEASONAL AND PERENNIAL.

Ephemeral vegetation, which constitutes the bulk of vegetation of the area, appear above ground just after first showers and complete their life cycle in incredibly short time. They die out as soon as soil gets dry. One of such pioneers in edaphic habitats is *Indigofera cordifolia.*

Seasonal vegetation in arid Rajasthan are those, which complete their life cycle in one season. The life cycle of this kind of vegetation begins with advent of first shower of rain. Degeneration and mortality starts with the onset of early winter season. *Tubulus terrestlis, Vernonja dnerea, Dicanthium annulatum* are predominant seasonal vegetation.

Perennial vegetation are species which grow throughout the year. Like seasonal vegetation, just after few showers of rainfall during the later part of June seedlings of these perennials, mainly grasses emerge from the ground. *Bothriclova pertusa* and *Cynodon dactylon* are the most common perennials.

The study of vegetation type, its extent and density, change dynamics gathers immense importance in light of decreasing green cover and increased awareness in the field of managing vegetation cover. Such studies become more relevant in the areas where vegetation is sparse and its conservation and preservation is of utmost importance.

The study has been carried out using IRS-1C WiFS data of three seasons - January, May & October (Photograph 1).

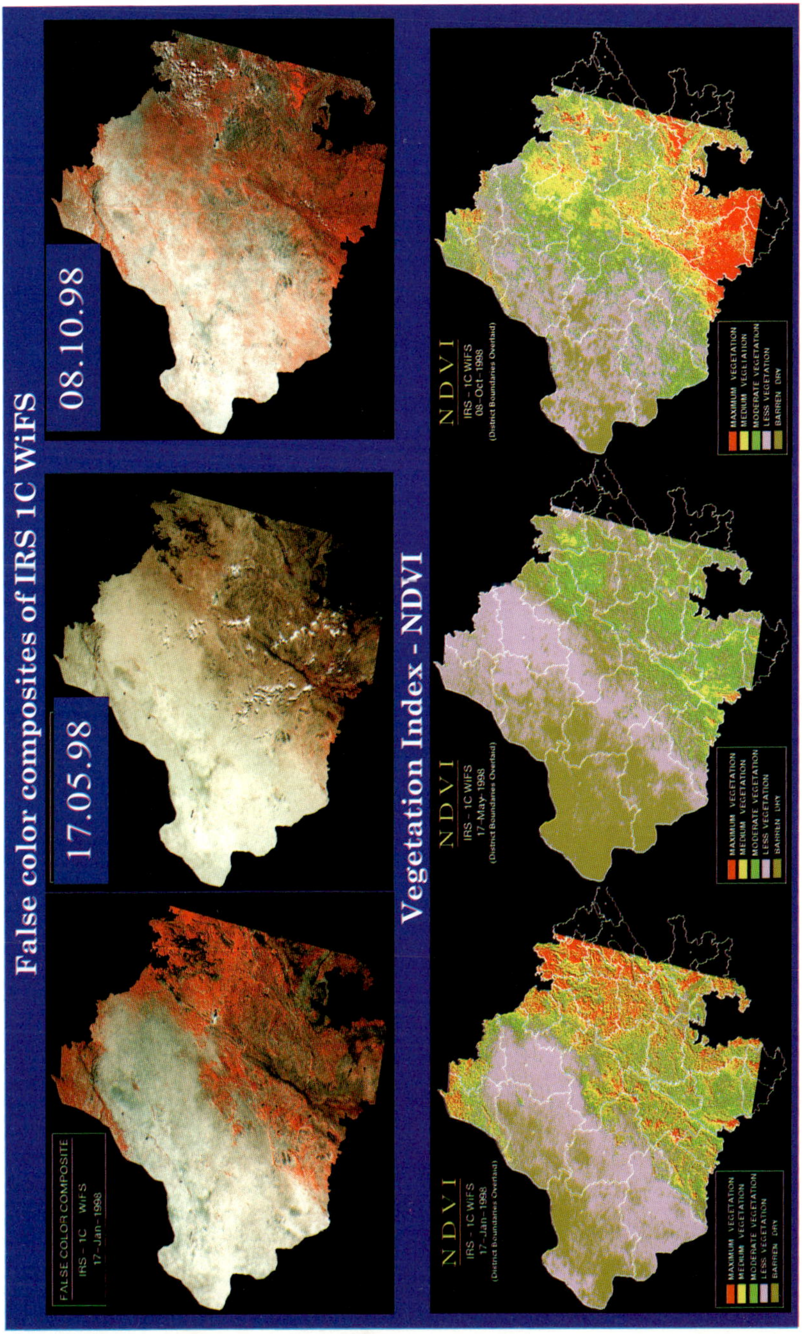
False color composites of IRS 1C WiFS
FALSE COLOR COMPOSITE
IRS - 1C WiFS
17-Jan-1998
17.05.98
08.10.98
Vegetation Index - NDVI
NDVI
IRS - 1C WiFS
17-Jan-1998
(District Boundaries Overlaid)
MAXIMUM VEGETATION
MEDIUM VEGETATION
MODERATE VEGETATION
LESS VEGETATION
BARREN DRY
NDVI
IRS - 1C WiFS
17-May-1998
(District Boundaries Overlaid)
MAXIMUM VEGETATION
MEDIUM VEGETATION
MODERATE VEGETATION
LESS VEGETATION
BARREN DRY
NDVI
IRS - 1C WiFS
08-Oct-1998
(District Boundaries Overlaid)
MAXIMUM VEGETATION
MEDIUM VEGETATION
MODERATE VEGETATION
LESS VEGETATION
BARREN DRY

Plate - 1

Objectives

1. To assess the density of vegetation.
2. Classification of vegetation by aggregation of three season data (January, May & October).
3. Dynamics of vegetation density. Correlation of vegetation density with cumulative rainfall data

Methodology

1. Grid base preparation for the study area
2. Geo-referencing of maps (1:250,000 scale) to the grid base
3. Geo-referencing of three season WiFS data and digitization & mask preparation
4. Vegetation Index (NDVI) of three seasons

Images obtained by space-borne imaging sensor integrate three components: a space component, atmospheric component and sensor component. Since, three season data has been used, it is required to normalize the data sets prior to the analysis. The digital numbers were converted to radiance values using bandwise saturation radiance and offsets and simple atmospheric corrections were performed by removal of Rayleigh path radiance. The NDVI has been generated, which shows strong near linear relationship to the fraction of visible radiation intercepted by vegetation canopy and appears to be almost insensitive to the variations in the canopy geometry. Each NDVI image was density sliced into five different classes viz. *Max. veg., Med. veg., Mod. veg, Less veg. And Barren/ dry.*

Classification of vegetation by aggregation

Density sliced images were recoded into two classes viz. vegetation and no vegetation. Aggregated output image was generated using lookup table depicting three vegetation types namely perennial, seasonal and ephemeral.

Dynamics of vegetation density

Dynamics of vegetation has been studied by analyzing three season images using lookup table. Negative change indicates the downward trend in vegetation viz. from maximum to medium or from moderate to less etc. Positive change indicates the upward trend in vegetation viz. from less to medium or moderate to medium etc.

Correlation of vegetation density with cumulative rainfall data

An attempt was made to correlate the vegetation development with rainfall using the NDVI of October season and rainfall data from June to September 1998. Rainfall data was collected from Central ground water board. The Cumulative rainfall was calculated districtwise from June to September as the NDVI image was of October 8, 1998 and considering that, the rainfall received till September might have had influence on vegetation development. Normalized mean NDVI was calculated for each district as a representative indicator of overall vegetation density.

Density of Perennial vegetation

During winter season vegetation density in three major categories i.e. max, med and moderate is about 25.68%. With the advent of summer heat the vegetation starts drying and it falls to 17.83%. Major change is observed in maximum density from 6.84% to 0.14% and medium density from 8.7% to 2.43%. This change can be attributed mainly to harvesting of rabi crop. After few showers during monsoon season, once again perennial vegetation mainly grasses emerge and density increases to 29.93%. This can be attributed mainly to kharif crop.

Density of Seasonal vegetation

Seasonal vegetation are mostly dominant after mansoon season 17.87% and during winter season 20.79%, whereas, during summer the density is almost insignificant 1.91%.

Dynamics of vegetation

39.68% area exhibit negative change from January to May season and positive change from May to October season. Majority of this area falls in eastern part of Rajasthan, some part in north and southern Rajasthan and majority of this area is under perennial vegetation. Decrease in density of vegetation is due to less rainfall between January to May, whereas, good rainfall during June to September results in vegetation development and hence indicates positive change.

22.94% area exhibit no change from January to May season and positive change from May to October season. Majority of this area is under Seasonal or Ephemerals vegetation. Positive change is attributed to rainfall and subsequent vegetation development.

12.96% area exhibit no change in both the season. This area is either predominantly sandy (in Western Rajasthan) or Rocky (in eastern Rajasthan)

Correlation of rainfall

As evident, In general, normalized mean NDVI values shows increasing trend with higher rainfall. Districts - Jaisalmer, Barmer, Bikaner, Churu and Nagaur exhibit lesser increase because these are mainly under sandy plain and crop growing season is limited to moisture availability period, which is less than 80 days. The ephemeral vegetation, which emerges soon after the rainfall complete their lifecycle within short time and also the seasonal vegetation, which is very sparse in these districts and yield less NDVI values.

Districts namely, Jaipur, Dausa and Alwar although having no limitation of physiography and having intensive agriculture areas show less NDVI values. This can be attributed to harvesting of crops (by month of October). The crop residues and grasses left behind yield low NDVI values.

Conclusion

The study demonstrates the vegetation in Rajasthan is edaphically controlled. Though, large part is covered by perennial and seasonal vegetation, but their density is highly dependent on the availability of rainwater and subsequent soil moisture. The study also demonstrates that the satellite RS can be effectively used for mapping and monitoring of constantly changing phenomena like vegetation at regional level because of higher revisit and larger swath. The temporal-spectral behavior of vegetation, which is related to phenological and climatic factors could be easily studied using high repeativity. The studies carried out for mapping of common property resources also proves the potential of remote sensing technology for the purpose.

REFERENCES

Bonner, D.F. 1978. Rangeland resources and world food needs. Proc. First International Rangeland Congress, Denver, Colorado. Pp. 5-7.

Dregne H.E. and Chou Nan-Ting, 1993. Global Desertification Dimension. In : Degradation and Restoration of Arid Lands (Dregne HE ed.).

International Centre for Arid and Semiarid Studies, Texas Tech University, Lubbock, Texas, USA, p.p 249-282.

Friedel, M.H. 1991. Range condition assessment and the concept of thresholds: a viewpoint. J. Range Manage. 44: 422-426.

Glantz M.H. and Orlovsky N., 1983. Desertification : A review of the concept. Desertification Control Bulletin 9 : 15-22.

Huete, A.R. and Jackson, R.D. 1987. Suitability of spectral indices for evaluating vegetation characteristics on arid lands. Remote Sens. Environ. 23: 213-232.

Nelson, R. 1988. Dryland management: The desertification problem. The World Bank; Environment Working paper No. 8.

Roundy, B.A. and C.A. Call 1988. Revegetation of arid and semiarid rangelands. In: P.T. Tueller, (ed.) Vegetation science applications for rangeland analysis and management. Kluwer Academic Publishers, Dordrecht, Holland.

Singh, P. and P.S. Pathak (Ed.) 1988. Rangelands: Resource and Management. Proceedings of the National Rangeland Symposium. Rangeland Management Society of India C/o IGFRI, Jhansi.

WRI 1993. World Resources 1994-95. World Resource Institute, OUP, Oxford, New York.

12

Multipurpose Trees as Alternate Feed Resources for Better Livestock Production in Dry Areas Particularly in Arid Zone

L.N. Harsh

Central Arid Zone Research Institute, Jodhpur - 342 003

Introduction

Tree based farming systems or agroforestry is an age old practice. In the recent past agroforestry has been recognized as a land-use system, which is capable of yielding both wood and food, and at the same time conserving and rehabilitating the ecosystem. Agroforestry has the capability to increase the productivity and at the same time maintain the nutrient balance as well as protect the nature. Hence trees in agriculture plays important role in sustainable agriculture as well as environmental protection.

However, in arid regions of India growing of trees in agricultural fields or in the pasture lands has a great role because due to frequent drought, agricultural crops failed and trees acts as crop insurance. During the drought period, trees provide leaf fodder for livestock, beside leaf fodder; they provide fuel wood, fence materials, fruits, gum etc. Even some times, farmers cut the trees and by sale of them, they earn the money for their livelihood. Therefore, in the twelve arid district of Rajasthan, animal farming becomes a major occupation. In India animal husbandry contributes about 32% of total agricultural revenue, which is 27% of GDP whereas in Rajasthan State it contribute 19% of the GDP. During the drought period, farmers lopped the trees and fed to their animals during the lean period and save their animals. Similarly farmers maintained trees in their orans/grazing lands for the providing feed to animals during lean period. During the monsoon, the in orans/grass lands, good quality of grasses available having 10-12% crude protein but as the monsoon reseed, the grasses dry and their crude protein fall to 4-5% which is not sufficient to maintain the animal health. Therefore, if green leaves/pods mixed with dry grasses, than total CP value of feed can be increased and sufficient to maintain the animal health.

Feed and fodder scenario

The fodder production vary widely depends on the cropping pattern, socio-economic conditions, and type of livestock. The cattle and buffaloes are normally fed on the fodder available from cultivated areas, supplemented to a small extent by harvested grasses and top feeds. The three sources of fodder suppliment are crop residues, cultivated fodder and fodder from common property resources, viz. forests, permanent pasture and grazing lands. The deficiency in field and fodder is identified as one of the major constraints in achieving desired level of livestock productivity. The shortage in dry fodder, green fodder and concentrates are 21.8, 61.5 and 47.1% compared with the requirements of 560,1006 and 79.4 million tones for current livestock population (2003) respectively. (Bhagmal *et al.,* 2006). However, the fodder availability scenario is ever worst in arid region. In the arid region there is a perennial shortage of 28 million tons/yr green fodders, which is 40% of the gross production. This may rise to 80-90% during extreme drought years requiring large volume of import of fodder from neighboring state.

Therefore, to bridge the gap between demands and supply if top feed species introduced than they can contribute 6 to 66% fodder to the system. Hence introduction of top feed species can bridge the fodder requirement as well protein requirements.

Fodder tree species

In arid region about 277 species reported (Bhandari 1999) of which 51 species of 34 genera belonging to 20 families are trees/ shrubs and have multipurpose uses (Table 1). Among them few are very commonly used.

As mentioned in Table 1, if these trees are properly explored/ optimally utilized for different purposes, then whole economy of western Rajasthan can be changed and economy of poor people can be improved.

The tree component is included in agroforestry system since ageses to get tree fodder especially during lean period in arid and semi-arid regions. Besides leaf fodder some other tree products are also used as a component of feed formulations and many of these are not properly evaluated. Since ageses people of the regions have screened the trees which have multiple uses and did not have any adverse effects on the companion crops/grasses and also does not have adverse effect on the human and animal health. The leaves, pods, fruits are being used as animal feed during lean period (summer) when under story grasses dried and their nutritive value declined and are not suitable to maintain the animal health. Therefore, if tree products are mixed with the grasses/crop stover than their nutritive value can be increased.

Table 1. Multipurpose trees and their uses

Sr. No.	Name of Tree/shrub		Uses
1	*Prosopis cineraria*	:	Leaf as fodder, green pods as vegetable, Dry pods as fruit, lops and tops as fuel, wood as minor timber and fuel, shed tree, nitrogen fixing tree
2	*Acacia senegal*	:	Leaf fodder, seeds as vegetable, wood as fuel, gum
3	*Capparis decidua*	:	Green fruits as vegetable, ripened fruits as fruit, Wood for making keel for floor chaki
4	*Tecomella undulata*	:	Wood for furniture
5	*Calligonum polygonoides*	:	Root as fuel wood, flower buds for vegetable, Green branches for camel fodder
6	*Salvadora oleoides*	:	Leaves as cattle feed, fruits for human consumption. Seeds for nonedible oil and cake for animals.
7	*Ziziphus nummularia*	:	Leaves as food, ripened fruits for human consumption, thorns for fencing
8	*Z. rotundifolia*	:	Leaves as food, fruits for human consumption, round wood as poles, thorns for fencing
9	*Ailanthus excelsa*	:	Leaves as fodder, wood for ply industries.
10.	*A. jaquimontii*	:	Twigs for making baskets, small quantity of gum, soil conservation etc.
11	*A. tortilis (Exotic)*	:	Used for sand dune stabilization, leaves and pods for livestock feed, gum, wood for furniture
12	*Prosopis juliflora (Exotic)*	:	Wood Used for charcoal making, pods as livestock feed, used for sand dune stabilization and reclamation of saline-sodic soils.

Fodder production and carrying capacity in silvi-pastoral systems: It revealed from the Table 2 that the reseeded pasture can maintain one ACU for approximately 310 days and if multipurpose trees grown along with the grasses than carrying capacity can be increased from 310 days to 344 days and 780 days in rainfed and irrigated areas respectively except in Bordi-pasture land which can maintain one ACU for 231 to 257 days. In irrigated areas stovers of bajra and wheat can maintain one ACU for 450 days and if *Ailanthus excelsa* planted on the farm boundaries (80 trees) than one ACU can be maintained for 780 days. Thus it revealed that introduction of perennials can increase the carrying capacity of land. Harsh *et al.* (1992) also reported that carrying capacity of sheep can be increased from 3.9 to 4.1 sheep/ha/yr in silvi-pastoral system *(Cenchrus ciliaris + Hardwickia binata, Colophospermum mopane)*

Protein Production in Silvi-pastoral Systems: As the rainfall reseeds from September onwards, the grass protein values also declined. At the pre- flowering stage, the protein value of grasses are very high (12%) but once it dried it declined to 4% and which is not able to main the animals. But in the silvi-pastoral systems tree leaves provides high protein ranged between 12-16%. Therefore, based on the protein requirement (1 gm/l kg of animal), it revealed that reseeded pasture can maintain one ACU for 333 days whereas silvi-pastoral can maintain one ACU for 567 days. In irrigated area this value can go upto 760 days (Table 2) the availability of crude protein to animals are higher when calculated on protein basis than total biomass production. In ardu based agroforestry system it was 1780 days/ ACU. Whereas on protein basis it will be 1760 days for one ACU.

Table 2. Forage production and crude protein yield in silvi-pastoral and pure pasture.

System	Tree density/ ha	Forage Production (Kg/ha)			Crude Protein Production (Kg/ha)		
		Grass	Trees	Total	Grass	Trees	Total
Cenchrus ciliaris Pasture land (Reseeded)	-	2500	-	2500 (313)	100	-	100 (333)**
C. ciliaris + Khejri	30	2250	450(20%)	2700 (337.0)	90	69.0 (57%)	155 (517)

C. ciliaris + Khejri	50	2000	750(37%)	2750 (344.0)	80	90 (65%)	170 (567)
C. *ciliaris + Bordi*	750	2000	53 (3%)	2053 (257)	80	6.5 (12%)	86.5 (288)
C. *ciliaris + Bordi*	1200	1750	109.0 (6%)	1850 (231.0)	70	13.0 (2%)	93.0 (277)
Wheat + Ardu (Stover)*	80	2000 (Stover) 1600	2400	6240 (780)	144	384 (126%)	528 (1760)

Protein requirement = 1 gm/l kg wt. of Animal.

Fig in parenthesis shows grazing days/ACU* based on biomass and available

Agroforestry Systems: Institute has developed many agroforestry systems and few of them described as under.

Acacia senegal: *A. senegal* grows widely in Western Rajasthan. It grows on sand dunes, sandy plains and rocky sites (Magra). Its density ranged between 10-30 trees/ha. Few trees are very old (>50 years). Local people collect the dry pods and extract the seed, which used for vegetable. Its seeds are one of the important components of Punch-Kuta. The seed production ranged between 1-5 kg/tree and sold in the market @ Rs. 20/ per kg. Hence it is not much economical. If this plant used for gum tapping as it has been practicing in Sudan and their economy depends on the gum export. Sudan contribute 80% of total gum export, whereas the same plants hardly produces 10-100 gm gum/tree in India. Therefore, in India gum-tapping techniques is to be developed for optimum gum production Khan and Harsh (1992) developed gum-tapping technique and by that technology on an average 500 gm gum/tree can be harvested. If 30 trees maintained at the farmland and used for gum tapping after 10 years of plantation than 20 kg/ha/yr gums can be harvested. If gum sold @ Rs. 200/ per kg than Rs. 2000/- can be earned. Farmers can additionally earn Rs. 2000 ha/year till the plants attained 20-25 years of age. Otherwise farmer was earning only Rs. 300/- to 600/ ha/year by sale of *A. senegal* seeds.

Under its canopy or in between the tree rows, farmers grow the crops as well as grasses. The tree leaves are palatable and browsed by Goat, Camel and Sheep. In Sudan also, farmers allows their animals to graze in the gum gardens before and after harvest of Gum from *A. senegal.*

Acacia tortilis: Acacia tortilis an exotic was introduced in India from Israel in early sixties and because of its fast growth and well adopted to arid regions it was planted on roadsides, sandy plains and sand dunes. On the sand dunes it grown very fast and able to check the sand erosion. Therefore, it was found to be one of the most suitable species for sand dune stabilization. It is very fast growing and yield 53 t/ha at the age of 15 years when planted at the spacing of 3 X 3 m. (Muthana and Arora 1980).

Besides, sand dune stabilization, it can be grown with the grasses i.e. in Silvi-pastoral systems. Muthana *et al.* (1985) reported that growing of *Acacia tortilis* along with *Cenchrus ciliaris* was more economical than growing grasses alone (Table 3). Besides gum production, clear bole can be used as timber in handicraft industries. Harsh (2006) reported that on an average 2 CFT wood/ tree can be harvested at the age of 15 years. If wood sold @ Rs. 200 per CFT than Rs. 60,000/- ha can be earned keeping 100 trees/ha (Table 4) It revealed from the Table 3 that cultivation of pure grass gives the return of Rs. 15,000/- per ha whereas growing *Cenchrus ciliaris* with *A. tortilis* return will be of Rs. 85,200/- per hectare.

Table 3. Economics from the silvi-pastoral components after 7 years of establishment

Treatment	Fuel yield (q/ha)	Grass yield (q/ha)	Revenue from fuel wood (Rs.)	Revenue from grass (Rs.)	Total Revenue (Rs.)
S_1G_0	60	-	3,000	-	3,000
S_2G_0	32	-	1,600	-	1,600
S_1G_1	50	55.8	2,500	1,395	3,895
S_2G_1	28	52.9	1,400	1,323	2,793
Grass without trees	-	46.0	-	1,150	1,150

***Salvadora oleoides* based silvi-pastoral system:** In the flood plains of the arid and semi-arid regions *Salvadora oleoides* is a common trees associated with grasses as well as crops. The density ranged between 10-30 trees/ha. In the oran lands in between the interspaces grasses are growing. Not only it grows in the orans/ grassing lands, it also grows in the farmer's fields. *Salvadora oleoides* does not have any negative effect on the crops as well as on the grasses, hardly, 10% reduction in crop was reported (Harsh *et al.*, 2005).

Table 4. Economics *Acacia tortilis* based system after value additions (After 15 yrs)

Treatments	Feed wood (Q/ha)	Grass yield (Q/ha)	Tree pods (Q/ha)	Timber CFT	Gum from trees	Revenue (Rs.)					Total Revenue (Rs.)
						Fuel wood @ 0/Q	Grass yield @100/Q	Tree pods @ 100/Q	Timber @ 200/CFT	Gum @ 100/kg	
SHG_0	120	-	12.6	400	163	7,200	-	12,600	80,000	16,300	1,02,060
S_2+G_0	100	-	16.5	300	140	6,000	-	1,650	60,000	14,000	81,650
SHG_1	100	100	10.5	350	150	6,000	11,000	1,050	70,000	15,000	1,03,050
S=+G_1	80	100	14.0	280	130	4,800	10,000	1,400	56,000	13,000	85,200
Grass	-	150	-	-	-	-	15,000	-	-	-	15,000

S_1= 200 trees/ha, S_2= 100 trees/ha, G_0= Without Grasses, G_1= With Grasses

The leaves are browsed by the Camel and Goat. The fruits are edible. The fruit production ranged between 30-80 kg/plant, but hardly 10% of produce used while rest of the fruit production perish. From the seeds about 50% non-edible oil can be extracted. Since its fruits are edible and to utilize its produce maximum, Harsh *et al* (2005), Khan *et al* (2004) developed the technologies for preparation of squash and Jam from the *Salvadora oleoides* fruits. Harsh et al (2005) reported that about Rs. 2400/- per ha can be earned more when *S. oleoides* grown with bajra (pearl millet) and if its fruit processed for squash and jam than Rs. 8574/- can be earned more than mono-cropping of bajra (Table 5).

Table 5. Gain per hectare when the farmer himself processes the fruits (300 kg)

Particular	Quantity Produced	Cost of Processing	Sale price of product	Gain	Total gain (Rs. ha^{-1})
Peelu squash	450 lit.	Rs.960 @ Rs,32/lit.	Rs.18,000 @ Rs.40/lit.	Rs.8400	Rs.8574
Fat from seed	9 kg	Rs. 90 @ Rs.10/kg	Rs.180@ Rs. 20/kg	Rs.90	-
Oil-cake	21 kg	-	Rs.84@ Rs.4/kg	Rs.84	-

***Ziziphus nummularia* based pasturelands:** Growing of *Z. nummularia* and *Z. rotundifolia* with agricultural crops/grasses are the age old practice in arid region of India. Two distinct example of silvi-pastoral were seen in Bhadria (Jaisalmer) and Desnokh (Bikaner) districts where more than 500 ha area is under silvi-pastoral system. In these area due to over grazing perennial grasses become completely vanish and only *Z. numularia* remained, but again due to excessive browsing by Goat and Camel they were in a poor state. Harsh and Shankarnarayan (1981) studied effect of Sheep and Goat or gazing on the well managed of *Z. numulalria* with *Cenchrus ciliaris* silvi-pastoral system. They have reported that in Silvi-pastoral system grazing by mixed flock of Sheep and Goat recommended for the proper utilization of grasses as well as bushes. Sheep grazed only grasses and bushes remained unutilized, similarly Goat browse only bushes and grassed become unutilized and if both mixed together, than both component utilized equally. They have further recommended that in a well developed silvi-pastoral system 6 sheep and 3 goat/ha can be grazed round the year. However, in the recent past efforts have been made to improve the *Z. nummularia Z. rotundifolia* by grafting-budding with improved variety of *Z. mauritiana* (cultivar Gola and Seb). If wild planted budded with Gola and Seb cultivar than within the same year good quality fruits can be harvested and from second year onwards 10-20 kg fruit/plant can be harvested. The high quality fruit prize ranged between Rs.5-l0 per kg. If 200 trees maintained along with the crops/grasses, than minimum Rs.15,000-20,000/ha can be earned by sale of ber only in addition of crop/grass production (Gaja *et al.* 2004).

In the recent past *Ailanthus excelsa* has planted along with the agricultural crops. It was initially planted as fodder tree, as its leaves are eaten by sheep and goat. But recently its wood used in ply wood industries. The market price of green wood is around Rs. 75/ql. In Sikar distt. more than 20 industries established based on *A. excelsa.*

The alternate land use system is not only drought proofing but improve the livelihood of the people.

REFRERENCES

Bhagmal, P.S. Pathak, V.S. Upadhyaya, J.N. Gupta, and G. Suresh 2006. Forage crops and Grasses. Forage crops and Grasses. In Hand book of Agriculture ICAR, New Delhi, pp 1128-1161.

Bhandari, M.M. 1990. *Flora of Indian Desert* (Revised edition), MPS Repros. Jodhpur

Gajja, B.L., P.R. Meghwal and L.N. Harsh, 2004. Economic performance of dryland Horticultural crop in Arid Zone of Rajasthan Ex Ante Analysis. *Cuvy Agric* 28 (1-2):59-62.

Harsh, L.N. and K.A. Shankarnarayan 1981. Animal production in Dhaman Bordi (*Cenchrus ziziphus*) grassland. Pp. 75-80. In Bordi (*Ziziphus nummularia*). A shrub of the Indian arid zone - Its role silvipasture. (Eds) H.S. Mann and S.K. Saxena, *CAZRL Monograph* No. 13.

Harsh, L.N., 2006. *Acacia tortilis* (Forsk) Hayne: a boon to arid region of India. In National Symposium on Livelihood Security and diversified farming system in arid regions of India. Pp.91. Jan. 14-16.2006.

Harsh, L.N., Hamid, A. Khan, P. R. Meghwal and A. Bohra, 2005. A study on economic potential of *Salvadora Oleoloides* Decne based agro-forestry system in Araba area, District Barmer. *J. of Non-Timber Forest Products* 12 (4) : 225 - 228.

Khan, H.A., and L.N. Harsh, 1992. Physico-chemical characteristics of Ethephon-induced Gum from *Acacia senegal. Indian Drugs,* 29(11): 507.

Khan, H.A., P.R. Meghwal, and L.N. Harsh, 2004. *Salvadora oleoides* Decne: An under utilized economic plants of the Desert. *Science Tech. Entrepreneur* 12 (12):61-63

Muthana, KD. and G.D. Arora, 1980. *Performance of Acacia tortilis* (Forsk) under different habitats of the Indian Arid Zone. *Ann. Arid Zone* 19 (1 and 2) : 110-118.

Muthana, K.D., S.K. Sharma and L.N. Harsh, 1985. Study of Silvi-pastoral System in arid zone. *My Forest* 21 (3) : 233-238.

13

Silvipastoral Approach to Combat Fodder Crises during Drought

M. Patidar

Central Arid Zone Research Institute, Jodhpur - 342 003

Introduction

The world food scenario is becoming more and more complex with the demand of food increasing at higher rate from a scarce resource. India with 2.2% of the global geographical area supports about 16% of total human and 20% of livestock population of the world. However, the productivity of the livestock in India is lowest in the world. The low productivity of Indian livestock is mainly due to chronic shortage of feeds and fodder coupled with poor quality. The majority of livestock are maintained in village grazing land or other waste lands which are in extremely degraded stock of vegetation and nutritive value. These problems are more acute in hot arid region, which occupies major part of north western India (28.57 million hectare) leaving western part of Rajasthan (19.6 million hectare) Gujarat (6.72 million/ha) and south we4stern part of Haryana and Punjab (2.75 million/ha). This region receives low rainfall and has high evaporation and high temperature with high wind velocity and low soil fertility. Rain fed agriculture combined with animal husbandry is the major basis of livelihood of desert dwellers. Frequent droughts result in failure of annual crops hence livestock enterprise is becoming more important in this region but availability of fodder is again bottleneck for livestock rearing the regions.

Livestock are valuable component of sustainable and economically viable farming systems of Rajasthan, endowed with huge livestock population which is at present 49.15 millions (livestock census 2003). Out of these, 23.62 million comprising 17.4% cattle, 13.55% buffaloes, 32.12% sheep and 35.39% goat are in arid districts of Rajasthan. The state contributes about 10% of milk, 30% of meat and 40% of wool production ink the country. Livestock sector in Rajasthan accounts 19% of state domestic products. The density of livestock except sheep and camel is less in arid region as compared to other parts of Rajasthan. Thus, are a per unit of livestock is more in arid region (1.82 ha/ACU) but fodder resources are much less than rest of the state. Inspite of large grazing area (about 53% of total land in arid region) of western Rajasthan, fodder availability is not sufficient to feed 21.92 million of livestock (Census, 2003). At present the requirement of total dry matter is estimated to about 39 million tones where as the availability is about 14-15 million tones of dry fodder and 8-9 million tones of green fodder under normal rainfall situation which is deficient to the extent of 40-50%. This may increase to 80-90% during extreme of drought years requiring large volume to import fodder from neighboring states.

The increasing deficiency of fodder is due to decreasing and deterioration of pasture and grazing land coupled with declined in cultivation of crops producing fodder. During normal year, the dual purpose crops contribute major portion of fodder but in drought conditions fodder production from dual purp-ose crops is reduced resulting fodder crises in arid zone of Rajasthan under such situation silivipastoral system is growing grasses in association with multipurpose trees/shrubs could help in combating the fodder crises during drought by providing nutritious green as well as dry fodder

Silvipasture and its importance

Traditionally silvipasture is used to grass farming in forest enclosures, which accommodate millions of livestock for grazing on the herbaceous growth. Extending this principal silvipasture is the conscious practice of cultivation of forage crops along with fodder trees shrubs some what similar to Agroforestry or taungya system. Although the word silvi mean tree and is new the practice of growing trees in the farm fields for their to feed value has been in vogue traditionally among the farmers of western Rajasthan. Their have encouraged and protected the population of *Khejri* (Prosopis cineraria) and *Bordi* (*Ziziphus nummularia*) in their crop fields to top them in winter for top feed and fuel wood. In simple world the term silvipasture denotes silvi means tree and pasture means grasses or grass + legume mixture but in real serve it is an ideal combination of grasses legumes and trees/shrubs for optimizing land productivity conserving plant soil and nutrient and producing forage, fuel, wood, timber etc. on sustainable basis. Thus this evolves re plantation substitution or intervention in the existing vegetation by desirable species (Deb Roy and Pathak, 1974). Recently Nair (1993) defined the silivipastoral system are land use system in which trees and or shrubs are combined with livestock and pasture for forage and fuel wood production on the same unit of land within this broad category several types of systems and practices can be identified depending on the role of tree/shrubs component viz, cut and carry system, live fences post browsing and grazing etc.

Silivipastoral system has special significance in arid and semi-arid region where there is levels vegetation and high rate of deforestation and land degradation due to low and erratic rainfall about 13% of land under potentially productive water lands a sustainable past of which can be reclaimed through silivipastoral system. This technology checks land degradation on one hand and provides much-recorded products viz. food, fiber, fodder and fuel wood, timber, medicine etc. on other hand.

Ruminants like buffaloes, cattle, goats and sheep are extremely important resources in the arid and semi arid regions of India, however, their productivity level by and large is extremely low because of low quality and seasonal nature of forage supply. The problem becomes more acute in dry seasons. In such situations, forage based silivipastoral system could be a promise. Besides, trees in silivipastoral system also provide shade to the grazing animals and serves as fences, hedge and wind breaks which intercept the wind blown particles and prevent soil erosion and conserve soil moisture silivipastoral system have potential to add organic matter to soil and improve soil structure of coarse sandy and alluvial land forms.

Suitable silvipastoral models for different land forms and rainfall situations

The selection of suitable species of trees/shrubs, grasses and legume for establishment of silivipastoral system depend on agro climatic condition of the regions. The following trees/shrubs and pasture species foe different land forms and rainfall zones are identified.

(A) Suitable species for silivipastoral system under different land form

Type of land	Suitable species		
	Grasses	Legumes	Trees/shrubs
Desert land and sand dunes	*Lasiurus sindicus Cenchrus ciliaris, Cenchrus setigerus*	*Lab lab purpureus, Clitoria ternatea, Atylosia scarabaeoides*	*Acacia tortilis, Acacia nilotica, Acacia senegal, Prosopis cineraria, Prosopis juliflora, Azadirachta indica, Ziziphus nummularia, Colophosperm-um mopane Dichrostachys nutan, Colligonum polygonoides*
Ravine lands	*Cenchrus ciliaris, Cenchrus setigerus, Dichanthium annulatum, Pannisetum pedicellatum, Saccharam spontaneum, Chrysopogon fulvus*	*Stylosanthes* spp, *Stizolobi-um decrigenum Macroptilium atropurpureum, Atylosia* spp	*Acacia tortilis, Acacia nilotica, Acacia catechu, Albizia lebbek, Albizia amara, Dalbergia sissoo Zizyphus spp., Ficus spp., Emblica officinalis, Eugenia jambolana*
Cultivable waste lands	*Cenchrus ciliaris, Cenchrus setigerus, Pennisetum polystachyon, Panicum antidotale, Sehima nervosum, Chrysopogon fulvus, Dischanthium annulatum*	*Macroptilium atropurpureum, Glycine jaranica, Clitoria ternatea*	*Albizia spp, Hardwickia binnata, Leucaena leucocephala, Acacia* spp. *Sesbania* spp., *Dichrostachys nutans, Prosopis cineraria, Ziziphus mauritiana*
Salt affected area	*Cynodon dactylon, Paspalium notatum, Chloris gayana, Lasiurus sindicus, Brachiaria mutica, Sporolobus marginatus, Urochloa*	*Glycine javanica, Macroptillum* spp. *Phaseolus junatea, Stylosanthes*	*Acacia tortilis, A. nilotica Prosopis juliflora, Salvadora spp, Ziziphus spp. Sesbania spp., Albizia amara Atriplex spp*

(B) Suitable species of grasses and trees on the basis of rainfall

Rainfall Zone	Grasses	Legumes	Trees/shrubs
150 to 300 m	*Lasiurus sindicus, Cenchrus ciliaris*	*Lab lab purpureus, Clitoria ternatea*	*Prosopis juliflora, Acacia tortilis, Acacia senegal, Ziziphus nummularia. Colligonum polygonoides, Dichrostachys nutans*
300 to 500mm	*Lasiurus sindicus, Cenchrus cilaris, Cenchrus setigerus,*	*Lab lab purpureus, Clitoria*	*Acacia. tortilis, A. nilotica, P. cineraria, Tecomella undulata, Colophospermum mopone, Ziziphus*

	Panicum antidotale	*ternatea*	*spp., Colligonum polygonoides. Azadirachta indica, H. binnata, Ailanthus exelsa, Albizia lebbek*
above 500 mm	*C. cilaris, Dichanthium annutatum, Chrysopogon fulvus, Sehima nervosum, Panicum antidotale, Heteropogon contortus*	Stylosanthes	*Acacia nilotica, Acacia catechu, Dalbergia sissoo, Leucaena leucocephala, Albizia spp., Alianthus excelsa, Hardwickia binata*

Forage Production of various Silvipastoral models

Forage yield from grasses grown in association with fodder-cum-fuel trees of various silvipastoral model varies depending upon soil and climatic condition. In traditional silvipastoral system, trees/shrubs are permitted to grow at random in fields on the boundary of the fields or community grazing land with grasses of low productivity and nutritive value. In arid zone of Rajasthan, Kaul and Ganguli (1963) recorded maximum combined yield of grass and Pala (1.0 t/ha) under 14% of land area covered with *Ziziphus* canopy. Similarly Ahuja (1980) recorded highest grass yield (1.1 to 1.5 t/ha) under *Khejri* and lowest (0.6 to 0.7 t/ha) under *Kumat* plantation of 14 to 18 year old with substantial contribution of *Cenchrus* and *Eleusine* species as under storey perennial grasses. The productivity and carrying capacity of grazing land can be increased through improved silvipastoral system. At Jhansi, Pathak *et al.* (1996) reported that the degraded lands producing hardly up to 1 t/ha/year biomass, which can be improved up to 10 t/ha through silvipastoral system. A study conducted at NRCAF, Jhansi revealed that an average total biomass yield of 12.62 t/ha/year recorded under silvipastoral system consisting of *A. amara* and *L. leucocephala* as trees, *D. cinerea* as shrub and *C. fulvus* + *S. hamata* as pasture was about 4 time higher than yield (3.16 t/ha/year) obtained from natural grassland (Rai *et al.,* 1999). Silvipastoral studies undertaken at CAZRI, Pali involving four tree species, (*Acacia tortilis, Azadirachta indica, Albizia lebbek* and *Holoptelia integrifolia*) and four grasses (*C. ciliaris, C. setigerus. D. annlatum and Panicum antidotale*) revealed non significance difference in the dry fodder yield under different tree species, however, the mean dry forage yield was maximum in *Dichanthium annulatum* (28 g/ha) followed by 25.1 g/ha in *Cenchrus ciliaris* and 22 g/ha in *Panicum antidotale* (Muthana and Shankarnarayan,1978). Silvipastoral studies at Jodhpur indicated that sowing of *Cenchrus ciliaris* under the cover *H. integrifolia* did not interfere with normal growth with an additional dry forage yield of 7 to 12 q/ha. In another study, Harsh *et al.* (1992) reported that a mixture of *C. fulvus* and *Sehima nervosum* grass produced 3.25 t/ha forage under canopy of five year old *Hardwickia binata*. Similarly Patidar (2006) recorded maximum herbage yield with *L. sindicus* + cowpea (90 and 31 q/ha) green and dry forage yield) followed by sole 25 (89 and 30.8 q/ha) green and dry fodder yield and *C. ciliaris* + cowpea system (84.6 and 28.6 q/ha green and dry forage yield grown in association with *C. mopane* and H binnata. The forage yield was not different due to tree species, however, application of 40 kg N/ha enhanced fodder yield by 13-14% over control. Yadav and Poonia (2003) recorded maximum biomass yield of grasses and tree from silvipastoral system with *Dichrostachys nutans* followed by *Prosopis cineraria* at existing stock of trees at farmers' field in arid climatic condition of Rajasthan. In a hort-pastoral studies on sandy rain fed of Rajasthan, Sharma and Diwakar (1989) reported that *Cenchrus ciliaris* in association with *Ziziphus mauritiana* system produced 1.2 t/ha forage yield and did not affect the fruit yield. Similarly Vashishtha (1995) recorded an additional forage yield of 4.7, 12.9 and 32.9 q/ha in first second and third year, respectively of planting of *Cenchrus ciliaris* in inter rows of *Ziziphus rotundifolia*.

Leaf fodder production

Studies on lopping intensity (Bhimya *et al.*, 1964) revealed that heavy intensity of lopping adversely affected the growth of *Khejri*. Further recurrent lopping reduced that leaf fodder yield irrespective of lopping perhaps due to successive reduction in the overall surface of new shoots. Need for a period of rest between two loppings of *Khejri* is, therefore, prima facie obvious for sustained yield of *loong* (leaf fodder of *Khejri*). It has been suggested that in arid area lopping of *Acacia* spp., particularly *A. nilotica* (ex. *A arabica*) should be regulated on a cycle of 4 years with restriction on lopping of thicker (diameter of 1.9 cm and above) branches. In Madhya Pradesh and Tamil Nadu lopping of *Hardwickia* spp was permissible only in times of scarcity.

Sharma and Gupta (1981) studied the effect of seasonal lopping on the top feed production and growth of *P. cineraria*. They observed that winter was the ideal season for lopping as it improved plant height and bole diameter of the tree as compared to trees lopped in rest three seasons i.e. spring, summer and monsoon. In similar studies on harvest of *Ziziphus nummularia* bushes in different seasons, the maximum top feed (*Pala*) 292 g/plant and shoot biomass 1170 g/plant was recorded in winter harvest as compared to other seasons (Saxena and Sharma, 1981). Fully grown (over 30 years of age) *P. cineraria* with well spread crown is reported (Muthana, 1980) to produce 25 kg air dried leaves, 5 kg pods, and 2 kg seeds in one year in 300 to 400 mm rainfall zone and a moderate sized plant (20 to 30) years age group) yields one kg seed. Complete lopping is reported (Kaul and Gyan Chand, 1977) to yield higher (58 to 72 kg/tree) than the lopping of the lower two third (28.48 kg/tree) and the lower one third (19.73 kg/tree) of the crown. There was no significant difference between two third and one third lopping treatments (Bhimaya *et al.*,1964). Top feed production from *G. tenax* under wide and close spacing ranged from 0.03 to 0.11 t/ha from without pasture plots and 0.03 to 0.09 t/ha from pasture plots similarly top feed from *Z. nummularia* varied from 0.03 to 0.04 t/ha. There were not much variation in top feed yield from close and wider spacing plots (Sharma et al., 1994).

Fuel yield in silvipastoral system

Afforested dunes in western Rajasthan produced 15 to 20 t/ha firewood in 200 to 300 mm rainfall zone (Bhimaya *et al.*, 1960). The firewood yield from *Prosopis juliflora* plantation was estimated at 15 t/ha at the end of 5 years (Kaul and Ganguli, 1964). Wide variations in yield with respect to the age and habitats of *P. juliflora* were noticed (Bhimaya *et al.*, 1967). Differences in the fuel yield between the habitats generally followed the pattern of rainfall i.e. increase towards the east.

Studies on fuel yield of *P. cineraria* revealed that for a diameter range of 12.4 to 25.7 cm the fuel yield ranged from 85 to 479 kg per tree with an average yield of 230 kg per tree of 20 cm mean collar diameter (Kaul and Jain, 1967). Kaul (1965) reported an average yield of 19 kg per tree firewood from *Calligonum polygonoides* at maturity (7 years old). Yield of 100 to 110 t/ha have been obtained from Casuarina equisetifolia raised in coastal Andhra Pradesh and managed under clear felling of plants with rotation age varying from 7 to 10 years (Rao, 1967). Away from the coastal area where water table and humidity were usually low, the yield varied from 25 to 65 t/ha (Ghosh, 1977).

SILVIPASTORAL SYSTEM FOR ARID REGION

Components : Grasses (*Cenchrus ciliaris* and *Lasiurus sindicus*),
: Legumes (Cowpea/ Lab lab)
: Trees (*Colophospermum mopane* and *Hardwickia binata*)

Production potential : About 40-50 quintal of green fodder, 1-2 quintal of tree leaves and 2-3 quintal of fuel woods.

Uses : Support 4 cattle units for 65- 70 days

It can be concluded from above discussion that over all biomass level of pastures, on-farm and other grazing lands could be enhanced by incorporating compatible species of trees, shrubs and grasses with proper management even under unfavorable climatic, edaphic and economic conditions in arid and semi-arid regions ultimately it will help in combating fodder crises during drought situations. This system also fulfill other requirements of farmers by increasing land productivity and resource use efficiency and help in minimizing soil and environmental hazard.

REFERENCES

Ahuja L.D. (1980). Grass production under Khejri. In *Khejri in the India desert its role in agro forestry* (Eds. H.S. Mann and S.K. Saxena). Monograph No. 11, CAZRI, Jodhpur pp 28-30.

Bhimaya, C.P. and Kaul, R.N. (1960). Some aforestation problems and research needs in relation to erosion control in arid and semi arid parts of Rajasthan. *Indian Forester* 80: 453-468.

Bhimaya, C.P., Kaul, R.N. and Ganguli B.N. (1964). Studies on lopping intensities of *Prosopis cineraria*. *Indian Forester* 90(1): 19-23.

Deb Roy, R and Pathak, P.S. (1974). Silvipastoral management. Indian farming. 24:41-42

Ghosh, R.C. (1977). Handbook on Afforestation Techniques. Controller of Publication, Govt. of India, New Delhi.

Harsh, L.N., Tewari, J.C., Burman O. and Sharma S.K. (1992). Agroforestry in arid regions. *Indian Farming* 40:32-37.

Kaul R.N. and Ganguli B.N. (1963). Fodder potential of Ziziphus in the shrubs grazing lands of arid zones. *Indian forester* 89:623-630.

Kaul, R.N. (1965). Silvical characteristics of *Calligonum polygonoides* I.a, Yaaron 15: 1-8.

Kaul, R.N. and Ganguli, B.N. (1964). Afforestation studies in the arid zone of India. Proc. Symp. *On Problems of Indian Arid Zone*, CAZRI, Jodhpur, India.

Kaul, R.N. and Gyanchand (1977). Forest tree plantation in arid zones. In *Desertification and its Control*, ICAR, New Delhi, pp. 169-202.

Kaul, R.N. and Jain, M.B. (1967). Growth attributes and their relation to fuel yield in *P. cineraria* (Linn). Mac. Bride (*P. specigera* Linn.), *Common wealth Forestry Review* 46 : No. 128.

Muthana, K.D. (1980). Silvicultural aspects of Khejri. In *Khejri (Prosopis cineraria) in the Indian desert – its role in Agroforestry* (Eds H.S.

Mann and S.K. Saxena). Monograph No. 11, CAZRI, JODHPUR, India, pp. 21 -24.

Muthana, K.D. and Shankarnarayan K.A. (1978). Scope of Silvipastoral system in arid region. In *souvenir on arid zone research silver jubilee*, CAZRI, Jodhpur, India.

Nair, P.K.R. (1993). An introduction to Agroforestry. Kulwar. Academic Publisher, The Netherland.

Pathak, P.S., Gupta, S.K. and Singh, P. (1996) IGFRI Approaches : *Rehabilitation of degraded lands*. IGFRI, Jhansi, 1-23.

Patidar, M. (2006). Strip cropping of grasses and legume in Silvipastoral system. In *livelihood security and diversified farming system for arid region* organized by Arid Zone Association of India held at CAZRI, Jodhpur during Jan 14-17, 2006, pp 33.

Rai, P., Solanki, K.R. and Rao G.R. 1999. Silvipasture research in India. *A review* 1: 107-120.

Rao, E. Venugopal (1967). Economics of plantations particularly of *Casuarina* and *Eucalyptus* with special reference to the sandy soils of Nellore, South Division of Andhra Pradesh Proc. *Eleventh Soil Conf.*, FRI, Dehradun, India.

Saxena, S.K. and Sharma, S.K. (1981). Production of a desert shrub (*Ziziphus nummularia*). In *Bordi (Ziziphus nummularia) – A shrub of the Indian arid zone, its role in silvipasture* (Eds. H.S. Mann and S.K. Saxena), CAZRI Monograph No. 13, pp 25-30.

Sharma S.K,. Singh, R.S., Tewari, J.C. and Uday Burman 1994. Silvipastoral studies in arid and semi and degraded lands of western Rajasthan. In *Agroforestry systems for degraded lands* (Eds. Panjab Singh, P.S. Pathak, M.M. Roy), Vol. 2, Oxford and IBH Publishing Co. Pvt. Ltd., New Delhi, 749-754.

Sharma, S.K. and Diwakar G.D. (1989). Economic evaluation of hortipastoral system in arid region of west Rajasthan. *Indian Journal of Range Management* 10(2):119-1222.

Sharma, S.K. and Gupta, R.K. 1981. Effect of seasonal lopping on the top feed (loong) production and growth of *Prosopis cineraria* (Linn.) Druce. *Indian J. Forestry* 4(4): 253-255.

Yadav G.L. and Poonia T.C. (2003) Production potential of grasses between existing stock of trees under Silvipastoral system in semi arid watershed area. In *national symposium on sustainability advancement and future thrust area of research on forage,* CCS, HAU, Hisar March 5-6, 2003 pp 53.

14

Propagation and Management of Grasses of Arid Zone

M.P. Rajora

Central Arid Zone Research Institute, Jodhpur - 342 003

Perennial pasture grasses provide feed to the livestock as well as help in soil conservation and increase soil fertility. In their princ-ipal role as livestock feeds, grasses in the tropics stand as the highest yielder of starch and proteins being the dominant compone-nt of tropical pastures, as the cheapest source of animal feed. Devel-opment of pastures in the arid areas of Rajasthan may be a good pr-oposition in amelioration of the economy of the farmers.

Rangeland and Pasture

The term 'range' refers to a vast area supporting natural vegetation, which is suitable for grazing and browsing by livestock. These lands are generally not suitable for cropping due to some soil limitations and moisture deficiency. Most of these lands are marginal and are in seriously degraded conditions on account of over exploitation.

Distinct from rangelands, "Pastures" are usually fenced and comprise relatively smaller areas of the better lands (land use class IV and lower). These pasture lands compete directly with cropping lands, where rainfall is higher than 350 mm annually.

Important Pasture Grasses of Arid Zone

Sewan (*Lasiurus sindicus* Henr.): Sewan grass is one of the dominant species of *Dichanthium-Cenchrus–Lasiurus* type grass cover of India. It survives under extreme arid conditions of Rajasthan and gives high forage yield. The grass is extremely drought resistant and thrives even in very low rainfall regions receiving 100-300 mm annually and has moderate tolerance to salinity. The grass comes well in sandy soils. It is highly nutritive and palatable to the livestock. Tharparkar breed of cattle of western Rajasthan is dependent on this grass. Under natural conditions crude protein (CP) ranges from 5.9 to 6.7 per cent but when cultivated and fertilised with NPK, CP can reach up to 15 per cent. Three hectare of *L. sindicus* is sufficient to support one adult animal (cattle) throughout the year. From good condition grasslands 50 to 75 q dry matter can be harvested from one ha. Seed production depends on various factors and average productivity ranges from 5 to 40 kg ha^{-1}.

Buffel grass (*Cenchrus ciliaris* L.): Commonly known as 'dhaman' in Rajasthan and 'anjan' in other parts of India, is considered as a very drought resistant species. It is adapted to a wide range of soil and climatic conditions and can be cultivated in areas receiving rainfall from 150 to 1250 mm annually. The grass has been successfully established on dry sandy to stony soils in arid and sandy-to-sandy loam in semi-arid parts of India. It can also grow on deep basaltic soils and on red lateritic soils. It is very palatable when young, and remains fairly palatable at maturity. It is one of the best forage grasses for semi-arid areas in the sub-tropics and tropics. Crude protein percent varies from 6 to 10. The varieties like Marwar Anjan, Bundel Anjan, CAZRI 358, Biloela and Molopo give 40 to 45 q ha^{-1} dry matter. Depending on the soil moisture 3 to 4 cuts can be obtained from August to April. Average seed productivity is 10-60 kg ha^{-1} although the potential is very high.

Birdwood Grass (*Cenchrus setigerus* Vahl): It is commonly known as 'moda dhaman' in Rajasthan and birdwood grass in USA. It is extremely tolerant to heat and drought and will grow in areas of annual rainfall as low as 200 mm making it excellent for improvement of low rainfall grazing lands. It hardly respond to winter rains. It is palatable to all kinds of livestock and can be propagated from seeds and may be used to renovate the denuded pastures or establishment of sown pasture. The grass is either grazed or cut and fed to livestock *ed lib* or turned into hay. Crude protein content of dry matter varies from 5 to 12%. High yielding varieties of *C. setigerus* produce about 40 to 50 q/ha^{-1} of green forage from 2 to 3 cutting during August to April in arid regions. The yield becomes double in semi-arid regions. The dry matter varies from 4 to 20 q ha^{-1}.

Karad (*Dichanthium annulatum* (Forsk) Stapf.): It is commonly known as 'karad' in Rajasthan and Marval, Palwan and Zinjoo elsewhere. In Rajasthan, it occurs mainly in the rainfall regions of $\geq$ 350 mm. It evades or endures drought well (Whyte, 1968). The plant grows to a height of 75 cm at maturity stage. It is regarded as an excellent fodder plant and highly valued pasture of high quality, vigour and productivity. An average hay production of 33-q ha^{-1} can be expected from a good *D. annulatum* stand (Dabadaghao and Shankarnarayan, 1973).

Blue Panic (*Panicum antidotale* Retz.): Blue panic is popularly known as 'gramna' in Rajasthan. It is often found on sand dunes and excellent sand binder and drought resistant perennial grass adapted to a variety of soils and climatic conditions. It is a very productive grass that can serve as a good source of fodder supply either as hay or silage. On the livestock farms of the tropical and sub-tropical countries this can be a good fodder source through out the year because of its wider adaptability, persistency, high productivity and good seeding capacity. It can be either grazed or cut and fed to the livestock. It shows that the grass is highly nutritive when cut at pre flowering stage. The grass is palatable to all kinds of livestock. However, it is not liked in mature stage as the stem becomes hard and woody.

Murat (*Panicum turgidum* Forsk.): Grown on sand dunes and sandy plains in open rangelands and cultivated fields of Thar Desert of Rajasthan and Gujarat. A common and characteristic species, often playing an important role in colonizing shifting sand dunes under 100-250 mm annual rainfall. Green fodder productivity ranges from 50 to 75-q ha^{-1} giving 20 to 30 q ha^{-1} dry matter. Although hard and woody, young stems serve as an excellent nutritive camel fodder. It has 59% dry matter at flowering and CP 4.9%.

Pasture Establishment and Improvement Techniques

Routine pasture improvement practices as applied under higher and more reliable rainfall conditions are out of question in the arid and semi-arid regions in northwestern India. Climatic conditions in these areas are too erratic and individual falls of rain are too ineffective for regular germination and continued early growth of introduced species. Introduction of new species is risky in the areas receiving average annual rainfall below 300 mm. While selecting the species priority should be given to the indigenous species. Following are the techniques, which may be helpful in getting higher production from pasturelands.

Fencing: Fencing of the area is essential to minimize the loss owing to overgrazing in pastureland. Iron angle posts with barbed wire are the most effective but the initial cost is high. Hence, ditch and core wall fencing may be constructed through 'sharmdan' or participatory approach. Top feed species like *Acacia senegal*, *A. tortilis*, *A. nilotica* var. *cuprisiformis*, *Calligonum polygonoides*, *Capparis decidua*, *Ziziphus nummularia* etc can be planted on the core wall.

Choice of species: For the less than 300 mm rainfall zone in deep sandy soils, *L. sindicus* is the only choice. On sandy loam soils, *Cenchrus* species are recommended. Above the 300 mm rainfall zone *C. ciliaris*, *C. setigerus*, *Panicum antidotale* and *P. turgidum* grasses are recommended. In heavier soils of the higher rainfall zone, *D. annulatum* and *Cenchrus* species are recommended.

Land preparation

On plain sites: Complete land preparation and sowing in rows at a distance of approximately 75 cm in less than 300 mm rainfall zone. The distance between rows should be 50 cm in areas of above 300 mm rainfall.

On slopping sites: In areas of above 3% slopes, contour, furrows and bunds are recommended. On still steeper slopes, closer distance of furrows should be adopted.

Propagation of Grasses

Sowing method: The sowing can be done by the following methods.

1. ***Direct seeding***: Pasture seeds are light and remain stuck to each other resulting in uneven distribution of seed and higher seed requirement. Therefore, the seeds are mixed with wet field soil in 1:5 ratio (by volume). This mixture is drilled in open furrows after effective showers of monsoon. Line sowing is preferred as it makes the after sowing operations easy.

2. ***Transplanting method***: Rooted slips/newly raised seedlings are planted at the onset of monsoon. Two to three seedlings are placed in a hill. For better establishment water is must at the time of plantation. After planting, 2-3 watering is also required.
3. ***Pelleted seed sowing***: Pastures are generally established through direct seeding or transplanting. Some practical problems are faced with direct seeding in general and for undulated areas in particular and transplanting being labour oriented becomes a costly affair. Hence, pelleted seed sowing may be an alternative for undulated terrain. Pellets are prepared with 100-125 g seed, 3-3.5 kg clay soil, 250 g sand and 250 g FYM. For one hectare 60-80 kg pellets are required.

Sowing depth and spacing

Caryposes of pasture seeds are small and should not be sown deep otherwise seedlings will fail to emerge at the surface. Seeds of the desert grasses have shown optimum germination under field conditions when sown at shallow depth of 1-2 cm with a light soil cover.

Cenchrus ciliaris, *C. setigerus* and *D. annulatum* with 50-75 cm row-to-row spacing proved to be the best for optimum forage production in arid zone. *Lasiurus sindicus*, *P. antidotale* and *P. turgidum* exhibit more radial expansion. The row-to-row spacing for these grasses vary from 75 to 100 cm depending upon soil types and rainfall conditions.

Sowing time and seed rate

The time of sowing is limited to period in which soil moisture and temperature are sufficient high to permit rapid germination and establishment. All the above-mentioned grasses can be sown after first effective showers of summer monsoon (last week of June to July).

The seed rate for grasses vary considerably, however, the seed requirement for one hectare is 6-7 kg for *L. sindicus*, 5-6 kg for *C. ciliaris* and *C. setigerus*, 2-3 kg for *D. annulatum*, *P. antidotale* and *P. turgidum*.

Pasture Utilization

Scientific utilization enhances the productivity and longevity of the pastureland. Pasture can be utilized by following methods:

Cut and carry method: Cutting at right stage is must. In a normal rainfall year two to three cut can be taken if cut at 50% flowering. It is essential to protect pasture in the year of establishment from grazing by animals. In the establishment year the fodder from the pasture may be harvested and conserved as hay for utilization in lean period. Higher production of pasture starts from the second year onward, hence utilization of pasture in rational way is advocated for effective regeneration in subsequent years.

Grazing: To harness the maximum productivity of pasturelands, controlled grazing based on carrying capacity is the best way. Suitable grazing system also enhances the longevity of the grasses.

Hay making: Excess fodder of good rainfall year can be converted into 'hay' for scarce years. Fodder banks may be created at least at Panchayat Samity level.

Production

Production of pasturelands depends on climate, soil type and management. The average dry matter productivity of *C. ciliaris*, *C. setigerus* and *L. sindicus* is 30-35, 20-25 and 35-40 q/hectare, respectively. The peak forage production from the pastures of *Cenchrus* species can be seen in 3rd and 4th year and starts declining after 5th year. Rejuvenation of pastures hence needs to be undertaken after 5 or 6 years. *Lasiurus sindicus* pastures last for decades.

Maintenance of Pastureland

Sowing of grasses in the abandoned and weed infested area often leads to failure of pasture establishment. Removal of weeds before seeding and subsequent weeding after sowing of grass seed would ensure better establishment of pasture and forage production. Two weeding after 3 and 7 weeks of germination are required during the year of establishment. Inter cultural operations by tractor drawn cultivator are required for better aeration. This provides favourable environment for root development ultimately resulting in increased productivity of pastures. Overgrazing invites less palatable annual grasses hence grazing should be as per carrying capacity of the pasture by following any scientific method. Pastures should be well protected from stray animals. If utilized through cut and carry system, cutting should be done at right stage. The remnant portion of grasses should also be removed before the start of dormant period. Reseeding should be done when the productivity start declining.

Prospects of Pasture Development in Arid Zone

The economy of the farmers of the Indian Thar Desert is based on animal husbandry. Pasture grasses are the common source of roughage to cattle. Improved pasturelands can provide manifold increased fodder over the existing ones and become a good source of income round the year. During droughts, when the crops fail, these grasses remain productive and provide fodder if only few showers are there. Improved pasture also provided employment to the rural people through out the year, i.e. selling the dairy products and forage in nearby areas, cities and other areas. The seed production of gasses may also be a beneficial enterprise.

REFERENCES

Dabadghao, P.M. and Shankarnarayan, K.A. 1973. *The Grass Cover of India*. ICAR, New Delhi.

Whyte, R.O. 1968. *Grasslands of the Monsoon*. London, Faber and Faber, pp. 325.

15

Drought Resistant Grasses and Weed Resources for Feeding Animals

Manjit Singh

Central Arid Zone Research Institute, Jodhpur - 342 003

Arid region of India experiences extremes of temperature (0 to 4°C in winter and 45-48°C in summer), low annual precipitation ranging from 450 mm in the east to 100 mm in the west, low humidity, high wind velocity and high evapo-transpiration. Soils are sandy, poor in nutrient with low water holding capacity and prone to erosion by wind and water. Natural vegetation in such edapho climatic conditions is sparse and stunted, predominantly spiny belonging mainly to *Dichanthium-Cenchrus-Lasiurus* grass cover type and very small area having *Sehima-Dichanthium* type.

The vegetation in deserts is a unique blend of small trees, shrubs and perennial grasses. In fact, the perennial component of these systems forms the lifeline and sustains humans and livestock, particularly during drought or near drought situations. Nearly half of the reported 682 species in hot arid zone of India are ephemeral and seasonal, the majority being monsoonal. Monsoon herbage has immense potential not only as livestock feed but also as desert food for human beings. In fact out of 682 species, 331 are ephemerals belonging to 44 families. Some 52 species are winter ephemerals; 17 occur in both monsoon and winter. In true sense, some 262 species (38.70% of flora) constitute monsoon herbage. These belong to 184 genera and 44 families. Amongst families, Poaceae has maximum i.e. 50 monsoon species followed by 25 in Fabaceae and 20 each in *Cyperaceae* and *Asteraceae,* 15 in *Euphorbiaceae,* 13 in *Convolvulaceae* and 11 in *Acanthaceae,* and remaining families having lesser than these species. Some 60 species appear dominantly during monsoon. These occur on all land uses i.e. croplands and grazing lands. These seasonal species germinate with onset of rains and complete their lifecycle by setting seeds with decline in soil moisture. These grasses and weeds are an important feed resource for the livestock of desert where pastoralism is dominant system especially in lower rainfall areas.

Seasonals that appear to be out of place in croplands are called weeds. Desert farmers know the utility of these seasonal as valuable forage. Quite often, these seasonal are collected, air-dried and stored in livestock yard for future use. A study by the Institute to assess their biomass revealed that weed biomass often far exceeded their companion crop i.e. their biomass was 2 to 4 times more than crop. These are mostly palatable weeds giving 655 to 1564 kg/ha of dry matter depending upon preponderance of weeds. Thus seasonals as weeds contribute immensely to livestock feed. The protein content varies from 10-21% (Patil *et al.*, 2009).

Monsoon herbage act as rain-gauges i.e. some species will germinate only after a particular amount of rain is received in one event. In contrast, there are species, which germinate at the very first event of rainfall irrespective of amount. These include large number of species of *Indigofera, Aristida* and *Cenchrus.* If there is no successive rain within a span of 15-20 days these will start flowering and set seeds as early as 20 days to 40-50 days. It has been found that at each germination event, 20-25% seeds of entire seed bank germinate and remaining are dormant. This is an ecological adaptation in annuals to germinate only a fraction of total seed bank so as to preserve the rest for next rain event. This ensures that if seedlings die or disappear without completing life cycle and formation of seeds, the remaining seed bank portion held with in soil is able to regenerate and continue the progeny. If however, there are rains they continue to grow vegetative till they face water stress. Growth up to 50-60 cm height has been observed in these species in case of prolonged wet spell. This indicates the plasticity in annual habit of these plants.

There are a large number of studies on estimation of dry matter yield of monsoonals and it can be intend that monsoon seasonal not only make up 34% of floristic composition, they also make up over 50% of biomass of grazing lands which is palatable too, in both protection and degraded conditions.

Nearly 70% of grazing land in Indian arid zone are degraded while other are in 14% fair, 13% good and 2-3% have excellent class (Shankar *et al.*, 1988). It is these 70% areas which have poor perennial plant cover and hence a preponderance of monsoon seasonals in these makes available grazable material @ 500 kg/ha to sustain livestock.

The composition of seasonal undergoes a drastic change upon grazing. A comparison of protected and grazed paddocks of grazing land at the end of 3 years and receiving 150 mm, 250 mm and 400 mm rain at Chandan, Beechwal and Palsana respectively revealed that grazed sites had preponderance of unpalatable annual forbs and such grasses which have awns that deter animals from grazing. Different studies indicate that same species behaves differently in two different rainfall situations.

With increasing grazing pressure in arid areas perennial like *L. sindicus* showed earlier flowering, fruiting and seed set. Compressing the vegetative phase is thus a mechanism of evading grazing pressure in this perennial grass. All the annual species showed opposite trend. With increasing intensity of grazing, vegetative phase was prolonged and seed setting was delayed in *Aristida funiculate, Indigofera cordifolia, Indigofera linifolia* and *Corchorus tridense.* This shift in phenophase could be related to platability. After the monsoon rains, annuals were eaten in preference over the perennials. With continued preferred removal of annuals, these palatables will be first to disappear due to overgrazing. It means that such shift in phenophases, caused by increase in grazing intensity, can effectively be used as indicator of beginning of deterioration of rangeland health and desertification.

Monsoon herbage is important component of natural grazing lands, pastures and croplands. They are the only plants available even after the perennials are removed. Persistence to withstand pressures and reappear once the favourable situations occur, make them highly resilient component of desert vegetation. Their life cycle strategy is also more suited to adverse temporal sequential stresses. Growth properly, they contribute up to 500-600 kg/ha dry matter in natural conditions. But there is a scope to capitalize its ecological properties to further increase its yield to double or treble by providing moisture just to flowering so that vegetative phase is prolonged.

In view of rich monsoon flora in our desert and the perennial vegetation persisting in the desert despite adverse climatic features, we are advantageously placed to use both annual and perennial grasses for livestock.

At present the focus of research is mainly on perennial grasses like *Cenchrus cillirus, C. setigerus* and *Lasiurus sindicus.* The varieties of pasture grasses have been developed at CAZRI, Jodhpur. The annual herbage is equally important and needs further study. Some of the perennial and annual grasses and pasture legumes commonly occurring in this area are as below:

Anjan: *Cenchrus cilliaris,* commonly known as dhaman in Rajasthan and anjan in various other parts of India, is a perennial grass with a stout root sock and is one of the dominant grass species in *Dichanthium-Cenchrus-Elusine* complex of grass cover of India. It generally grows in warm, dry sub-tropical countries. It has been considered as a highly drought resistant grass species that can grow in wide range of soil and climatic conditions and can be cultivated in areas receiving rainfall from 150 to 1250 mm annually. The Institute has released variety CAZRI-75.

Marwar Anjan (CAZRI-75) of Cenchrus ciliaris: It is a tall (90-125 cm), erect, thick stemmed, leafy, drought hardy, with green yield potential of 70 q ha/year yielding 53% higher over the check variety Molopo. It has high crude protein (8.3%) which is 32% higher over the check.

Dhaman: *Cenchrus setigerus,* commonly known as moda dhaman in Rajasthan (in contrast to Ru-dar dhaman for *C. ciliaris)* due to absence of cilia on the spikes, is a perennial, forming clumps with somewhat bulbous base and grows up to height of 50 cm or so. A high yielding variety developed at the institute is described here.

Marwar Dhaman (CAZRI-76) of Cenchrus setjeerus: It is a clonal selection from exotic material EC 17655. It is a medium tall (50-60 cm), thin stemmed, leafy (leaf-stem ratio, 2.3), foliage remains green upto December. It is drought hardy, high tillering ability (29 tillers plant^{-1}), fast regeneration (2-3 cuts in a year), average green fodder yield (40 q ha^{-1}) and 15 q dry matter yield which is 38% higher over the check var. Pus a Yellow Anjan.1t is also nutritious, having 9.6% crude protein at 50% flowering stage.

Sewan *(Lasiurus sindicus* Henr.) The grass is confined to sandy to sandy loam habitat where the rainfall is less than 350 mm annually. Sewan is also called as the 'King of desert grasses' due to its drought resistant characteristics. It grows round the year except harsh temperatures, depending on moisture availability. It is a tufted perennial grass, 100 cm tall from a more or less oblique and woody, rhizomatous rootstock. Sewan is a common grass often forming extensive pasturelands for considerable area in western Rajasthan.

CAZRI-30-5 of *Lasiurus sindicus:* The variety was developed through mutation breeding. It yields 97 q green fodder and 33 q dry matter yield per hectare which is 38 per cent higher over the check var. Jaisalmer Local. It attains a height of 1 m. It is drought hardy, remains productive even in the regions receiving an annual rainfall of less than 100 mm, leafy (leaf-stem ratio 1.9 which is 46% higher over check). Its crude protein yield is 3.75 q ha^{-1} which is 63 per cent higher over check. Its pasture remains productive up to 15 years in the arid zone.

Gramna *(Panicum antidotale* Retz.). It has a wider adaptability not exacting in soil/requirement. It is tall growing perennial grass, grows a height of 150-250 cm, has woody stem (on maturity) with a creeping rootstock giving off stolen and thickened modes. It is nutritive for livestock at flowering, grazed, used as art and carry method as well as hay.

Murtiogas *(Panicum turgidum* Forsk.). It is a grass of sand dunes and sandy plains in arid climate. A perennial, glabrous up to 1 m tall, roots fibres thick and felty. Excellent fodder particularly for camels.

Aristida plumosa: A dense tufted, perennial grass, up to 30 cm high. It grows in sandy soil and attains optimum growth during summer monsoon. It is a valuable fodder grass, and chief fodder for sheep.

Bhurat *(Cenchrus bjflorus* Roxb.) This is a monsoonal tufted, glabrous annual grass species. Comes well in sandy soils. It is the most common grass used very much in fodder scarcity, grazed as well as cut and carry system.

Gharadhob *(Eleusine compressa* Forsk.). It is a prostate, profusely branched and widely spreading perennial. This is a grass of gravely and sandy soil. Optimum time for growth is summer monsoon. It is a fairly good fodder for cattle.

Pasture Legumes

Bekrio *(Indigofera Cardifolia* Heyne.). Bekrio is a browse species found in open rocky slopes/gravels. A prostate diffuse or sub erect, annual herb, usually clothed with long, white modifixed hairs. Stem up to 30 cm long. A very good legume fodder for cattle and goats. It is found throughout the country.

Goilia or Jhil *(Indigofera oblongifolia* Forsk.). It is found on open dry places. It is a perennial species, an erect ashy-grey, twinggy shrub, 60-130 cm high. This is also a browse species. Goats and sheep eat this legume fodder.

Butterfly pea *(Clitorea ternatea* Linn.). This perennial pasture legumes comes well in sandy to deep alluvial and heavy black soils. It is a twining perennial with a few foliage branches distributed through tropics and subtropics. Stem fibrous or sparsely pubescent. Dry matter production ranges from 20 to 40 q/ha. Herbage contains 10 to 25% Crude protein. Grazed in pasturelands.

Low productivity of Indian livestock is a matter of concern, which is predominantly due to low quality and inadequate fodder and feed resources. Various natural and cultivated forages provide nearly 60% of the green fodder requirement of India's livestock. The challenge before us is to brig the gap between a demand of about 900 million tonnes and supply of 550 million tonnes.

REFERENCES

Patil, N.V., Mathur, B.K., Bohra, R.C. and Patidar, M. 2009. Nutritional analysis of various tree leaves, bushes, shrubs, grasses and weeds available for small ruminants in arid region. In : Forage symposium on Emergency trends in forage research and livestock production, CAZRI, RRS, Jaisalmer, Feb. 16-17, 2009, pp. 134.

Vinod Shankar, Suresh Kumar and R.K. Tyagi 1988. Grazing researchs of semi arid and arid regions. In Pasture and forage crops research, (Ed. Panjab Singh) Range Management Society of India. Pp. 63-85.

16

Feed and Management in Arid Region

Pratap Narain

Rajasthan Agricultural University, Bikaner - 334 004

Indian arid zone covers 31.8 lakh km^2 nearly to 10.16% of total geographical area. The hot region of nearly 62% lies in the state of Rajasthan covering 12 western districts of Barmer, Bikaner, Jaisalmer, Jodhpur Churn, Nagaur, Jalore Jhunjhunu, Sikar, Pali Hanumangarh and Ganganagar.

The region suffer from low and erratic annual precipitation (100-420 mm), high evapotranspiration (1500-2000mm) and poor physical and fertility condition of soil making crop farming a big gamble. Animal husbandry plays an important role in live hood security and economic sustenance of people. The arid region has about 28.57 million livestock population which during last century has increased by 61%. The man: livestock ratio is much higher in arid area and presently per 100 human population there area 124 livestock numbers in the area which may go to 200-300 is extreme districts of this zone indicating economic dependency of farmers on livestock husbandry (Narain and Rajora 2004).

In spite of harsh climate and feed scarcity, arid region is gifted with best breeds of cattle; Tharparkar, Rathi, Kankrej and Nagauri. Sheep: Chokla, Marwari, Magra, Jaisalmeri, Patanwadi, Nali and Pugal, Goats: Marwari, Kutchi and Camels: Bikaneri, Jaisalmeri and Kutchi.

Tharparkar cattle breed is native of Thar desert and reared primarily for milk production. Rathi, a milch cattle breed found widely among nomads who maintain these animals on ranges. The Kankrej and Nagaur are excellent drought breeds. The wool produced by arid sheep is of coarse and carpet type and annual production is about 1.5 to 2.0 kg whereas the goats often termed on poor man's cow has a distinct social, economic, managerial and biological advantages over other livestock in the region (Singh 2004).

The drought in the arid region is a common occurrence with frequency as high as 48 to 63 in past century and sometimes persistent of 3-6 years, which affect agricultural and fodder production. This coupled with degraded pastures due to overgrazing cause feed deficit for large population of animals.

In India there is shortage of 40% dry and 25% green fodder (FAO 1998). In Rajasthan the fodder deficit in western Rajasthan is estimated to be as high as 60% of the demand and might range form 55% in western Rajasthan districts of Bikaner, Jaisalmer and Barmer to 69% in the Central districts of Jodhpur, Nagaur and Churn and 72% in the eastern districts of Pali and Sikar. During drought years the overall deficit might go as high as 76% of the demand, and may range from 76% in fodder situation is better due to excellent pasture of *L. sindicus* western districts to 81% in the central districts and 82% in the eastern ones (Venkateswarlu et *al.* 1992). It is unavailable that inspite harsh climate situation in higher arid areas. The acute shortage in the central and eastern districts is on account of extensive cropping, fragmentation of land holdings, intensive cultivation and shrinking of pastures and grazing lands. The gap between the demand and supply will continue to widen if appropriate strategies are not initiated to enhance biomass and forage production substantially (Narain, 2000).

Nearly 1/3rd area of arid zone of Rajasthan are wastelands, of which 50% are grazing lands and 45% are sandy wastes (Balak Ram *et al.,* 2005). The production from the available grazing lands is hardly 300-400 kg/ha biomass. In the recent past, there has been a general awakening at various levels among land and stock owners regarding the need for an application of advanced technology to improve the deteriorated and depleted range pastures, in order to obtain a sustained level of production through management practices based on accepted principles of range management which include (1) Adoption climax key species in the range and (2) the concept of relative potentialities of individual species in a range grassland community from the point of view of animal grazing capacity and utilization.

The situation further warrants adoption of other feed management strategies for ensuring feed and forage availability from agriculture, forest and other areas during the droughts without further delay for supporting the livestock component of agriculture which remains base for livelihood of peasants of this region. Various approaches for ensuring feed and fodder security are discussed in this chapter.

1. Management of Grasslands

The grazing based animal agriculture is the predominant occupation and major source of livelihood of the people in arid regions. The rangelands occupy on an average 45% land and in extreme arid tracts like Jaisalmer 95% of the area is used as rangelands or grazing lands (Shankar and Kumar, 1987). Lands of arid Rajasthan are in depleted condition due to their over exploitation, abuse, encroachment, inadequate protection and absence of any grazing policy. These lands have potential to produce 4 to 5 times forages or even more by adopting following appropriate management practices.

Need of proper land use policy

Lands falling under Class V to Class VIII land capability classes are unfit for crop cultivation and those under class IV are subject to wind and water erosion. These lands should be utilized for pasture/ grassland development, and not forced under cultivation. Such policy will augment forage production and control land degradation. Areas receiving <500 mm average rainfall are cultivated mostly with pearl millet, cluster bean, mothbean and mungbean. In sandy soils with less than 200mm rainfall *Lasiurus sinclicus* has been found most adaptable and productive over all other grasses and which need be promoted.

Considering the fact that the relative palatability of a range species determines its intensity of use by the grazing livestock, observations recorded by Dhabadgao and Murwaha (1962) for five of the most important species of arid Western Rajasthan; and palatability rating done by comparing them with *Dichanthium annulatum,* grass considering its palatability 100 per cent indicate the importance of arid grass species for arid rangelands (Table 1 and 2 below).

Table 1. Relative palatability of six arid zone range grasses (for cattle)

	Pasture stage	Pre-flowering	Flowering stage	Ripe stage	Deap ripe stage
Dichanthium annulatum	100.0	100.0	100.0	10.0	100.0
Cenchrus ciliaris	51.9	169.7	52.8	125.0	266.7
Cenchrus setigerus	43.9	55.6	32.4	97.1	216.7
Lasiurus sindicus	43.2	31.5	37.9	109.7	125.0
Panicum antidotale	19.0	30.8	10.7	38.5	16.7
Panicum turgidum	49.7	14.0	31.2	59.4	33.3

Table 2. Average nutritional values of some range grasses

Range Plant Species	Percentage of dry matter yield			Total available digestable protein (Kg/hectare)
	Crude protein	Phosphorus	Digestible protein	
Cenchrus ciliaris	12.82	0.517	8.60	522.15
Cenchrus setigerus	8.77	0.602	4.89	168.87
Panicum antidotale	12.24	0.489	8.07	667.92
Dichanthium annulatum	5.05	0.482	1.48	19.05
Lasiurus sindicus	9.05	0.557	5.15	287.31

Protection of grasslands

Fencing of an area is essential to alleviate heavy livestock pressure. Barbed wire fencing is by far the most effective and economical in long run. But the initial cost of fencing is very high hence the farmers may construct ditch and core wall fencing through participatory approach. Border can be planted by species like *Acacia senegal, A. tortilis, A. nilotica* var. *cuprisiformis, Calligonum polygonoides* and *Cappairis decidua*. It has been proved that by mere protection, a lot of benefit in terms of vegetation availability can be obtained. After 2 years of protection, the forage yield is reported to increase by 148, 92 and 116% in 'poor', 'fair' and 'good' classes of rangelands, respectively in the tropical arid region, having sandy-loam soils (Paroda *et al.*, 1980). Similarly Cleaning of thorny bushes like *Lycium barbarum, Balanites aegyptica, Acacia leucophloea and Mimosa hamata,* should be done mechanically to improve grassland productivity.

Reseeding of grassland

Reseeding with high yielding perennial forage species suiting to arid tracts need be done for ensuring optimum grass production from the rangelands. It involves soil working, seeding of grasses and after for desired cover.

Recommended pasture grasses: Dichanthium annulatum on heavy soils with annual rainfall above 380 mm, *Cenchrus ciliaris* and *Cenchrus setigerus* on well-drained soils under medium to low rainfall, *Lasiurus sindicus* on sandy soils with low precipitation, *Panicum antidotale* on well-textured soils with annual rainfall of 250 mm and *Sehima nervosum* for hilly terrain are recommended grass species. The high yielding varieties of theses pasture grasses viz., CAZRI 75 of *C. ciliaris* and CAZRI 76 of *C. setigerus* are good varieties developed by CAZRI which can be utilized for reseeding.

An un-notified variety CAZRI 358 is doing well in almost all the habitats of arid zone. The variety has wider adaptability and can tolerate grazing pressure.

Soil Management involves land ploughing and unwanted vegetation be grubbed by the end of June. Sowing of mixture of seeds @ 5-7 kg ha is recommended to take care of rainfall aberrations and speed of germination. The grass seed should be mixed with 4-5 volumes of moist field soil and drilled uniformly 2-3 cm deep in lines 75 cm apart after the first effective rains and covered with light soil. Depending upon rainfall, dressing of 20-40 kg N/ha is recommended to reap higher forage production from reseeded grasslands.

Such reseeded grasslands need at least two weeding in first year of establishment. The reseeded grasslands should not be opened for grazing in the first year of establishment. In the second year and onward grazing should be so regulated that proper seed formation and shedding of seeds take place to guarantee satisfactory regeneration.

In order to improve productivity water conservation measures such as contour furrows, contour bunds and staggered contour trenches are extremely valuable. Reseeding of *Cenchrus* species in furrows revealed 130% increased production over control (Singh, 1992). Inter-row water harvesting system in which 30 cm wide ditch alternated with 70 cm wide raised bed was planted with two rows of pasture legume and grass on the edge of the ditch. This system of planting increased fodder yield of legumes and grasses by 124 and 66%, respectively over conventional flat bed system of planting (Singh, 1992).

Productivity potential of arid grasses

Well-established pastures become ready for cutting/grazing even after a fortnight of effective monsoon. The dry matter production of pasture grasses on demonstration plot was 30-35 q/ha for *C. ciliaris,* 20-25 q/ha for *C. setigerus,* 35-40 q/ha for *L. sindicus* and *P. antidotale.*

Cost: Benefit of reseeding of pasturelands

Cost of reseeding varies from Rs 3500-4000 ha^{-1} with a return of Rs 1000 to 2000 ha^{-1} in first year but higher returns in subsequent years with little inputs make the reseeding practice cost effective. Over a period span of five years benefit: cost ratio becomes as high as 3.13:1 (Table 3; Gajja, 2000).

Table 3. Benefit: cost ratio of pastures in arid zone of Rajasthan

Establishment year	Cost (Rs/ha)	Returns (Rs/ha)	B:C ratio
1 year	5050	1250	0.25:1
2 year	2000	2600	1.3:1
3 yea r	2000	5960	2.63:1
4 year	2000	6250	3.13:1
5 year	2000	6250	3.13:1

2. Pasture/grassland utilization

Considering low yield of pasture in arid region and high labour cost, harvesting is not a viable proposition unless mechanized harvesting and bailing is restored to, where sandy terrain imposes some difficulties. In case of reseeded, established and fertilized pastures, if forage yield are high, it is better to harvest 1/4th of the area at 50% flowering and preserve it as hay for feeding to milking stock, suckling, sick animals and working bullocks during adverse conditions. Preservation of surplus fodder can be brought about through hay and silage making, by treatment of roughages and bailing to blocks/rolls to facilitate transportation.

Controlled grazing: Most of the naturally growing grasses are very poor in productivity; the problem is further aggravated by grazing for excess to carrying capacity. Improvement of degraded grasslands need protection; controlled grazing and introduction of high yielding perennial pasture grasses and eradication of unpalatable perennial shrubs/grasses. Condition of grassland is very crucial for controlled grazing, which depends upon carrying capacity (Table 4). Grazing saves the cost of harvesting of poor uneconomical grassland and offers ground exercise to livestock. Livestock while grazing work up the soils, manures grazing lands, break the top crust of the soil, thereby encouraging better percolation of water and range production. Deferred rotational grazing is best practice advised for low rainfall regions but covered with high yielding nutritious *L. sindicus* grass spp. to increase number of grazing days, maintenance of proper vegetation through self seeding, fertility levels of soil and to minimize erosion.

Table 4. Carrying capacity of various classes of pasture

Condition class of pasture	No. of ACU (300 kg body wt.) per 100 ha	Productivity (q/ha)
Excellent	25-30	15
Good	20	10
Fair	17	7.5
Poor	13	5
Very Poor	6 and less	2

3. Top feed-cum-shade trees and shrubs on grasslands

Traditionally a few feed fodder cum shade trees are maintained on pastures to supplement nutritious fodder during lean period as well as to provide shade in hostile weather. These trees are heavily lopped, exploited and are in depleted poor state. *Prosopis cineraria,* which is extensively lopped in winter, can withstand recurrent lopping without causing any detriment at effect its growth or subsequent leaf yield. A moderate sized mature tree of the species yields about 15 kg of leaf forage locally known as '*Loong:* Z. *numularia* locally known as '*bordi*' is also heavily lopped in winter for its leaf fodder known as '*Pala:* Investigation at Pali and Jodhpur has shown that the species growing at medium density of 14% is optimum in natural grazing lands for maximum fodder production, which may yield about 125 kg of '*pala* 'per ha.

Some suitable top feed species on rangelands with systematic lopping should be integrated for good pasture management. Trees can also be planted along the boundaries, inspection path and approach roads to serve dual role of windbreaks and source of fodder. A good pasture may have 30 to 35 such trees per hectare. The important top-feed-cum-shade trees for rangelands are *Acacia nilotica, Prosopis cineraria, Salvadora oleoides, Acacia senegal, Albizia lebbeck, Anogeissus rotundifolia, Anogeissus pendula, Azadirachta indica, Grewia tenax,* G. *spinosa, Techomella undulata and Hardwickia binata.* Shrubs like Z. *nummularia, Calligonum polygonoides, Haloxylon salicornicum, Capparis decidua* and *Acacia jacquemontii* are also relished by goats, camels and livestock.

4. Silvipasture Systems for Fodder Production

Silvipastoral system is best suited for areas receiving <200 mm rainfall, or for degraded rocky-gravelly areas. Tree species highly compatible with grasses are: *A. senegal, A. tortilis, A. lebbeck, T. undulata, Colophospermum mopane, Dychrostachys nutans, Hardweckia binata, Z. nummularia* and *Z. rotundifolia.* Among the pasture legumes *Clitoria ternatea* and *Lablab purpureus* showed good compatibility with *L. sindicus* and *C. ciliaris.* The carrying capacity of a pure pasture was found to be 3.9 sheep per ha after 9 years of establishment, while that from a silvi-pastoral system was 8.5 sheep after 7 years. Under grazing study, *Z. nummularia* with grass strips in 1:2 ratio led to higher economic returns due to higher weight gain of the animals and higher wool production. Silvipastoral studies involving plantation of *A. tortilis, Azadirachta indica,* A. *lebbeck, Holoptelia integrifolia,* and *C. ciliaris, C. setigerus, D. annulatum and P. antidotale* revealed non-significant differences in grass yield at Pali, which was 2.8 t/ha for *D. annulatum,* 2.5 t/ha for *C. ciliaris* and 2.2 t/ha for *P. antidotale* (Muthana and Shankarnarayan 1978).

Growth and top feed production of *Z. nummularia* and *Z. rotundjfolia* on established pasture of *C. ciliaris* at Samdari showed that after five years of establishment, growth of the jujube species was more than two times and *'pala' production* was more than six times in control (no pasture) as compared to established pasture due to competition of grasses. However, the total forage production varied from 135 to 1269 kg/ha^{-1} and pasture yields were non significant with or without jujube. Planting of jujube in pasture is recommended at 6 m x 6 m spacing (Sharma and Vashishtha, 1985).

In an improved silvipasture, *H. binata* was taken as tree component at 3 m x 3m with *C. cilaris.* Results of nine years study (Table 5) revealed that average carrying capacity of the system was 4.1 sheep/ha/year against 3.7 for sole pasture and 1.6 for sale tree plantation (Narain and Tewari, 2005). Silvipastoral systems having a three-tier canopy, i.e. grass + legume (first tier), shrubs (second tier) and fodder-cum-fuel trees (third tier) must be given high priority for improving the feed availability in this region.

Table 5. Production potential of *Hardwickia binata* + *Cenchrus dlians* based improved Silvipasture

Year after planting	Grass yield (q/ha)	Leaf fodder yield (q/ha)	Total fodder yield (q/ha)	Carrying capacity (sheep/ha/yr)
1	8.2	-	8.2	1.9
2	12.3	-	12.3	2.9
3	16.4	-	16.4	3.9
4	20.5	-	20.5	4.9
5	16.4	-	16.4	3.9
6	16.4	-	16.4	3.9
7	12.4	23.1	35.5	8.5
8	8.4	6.6	15.0	3.6
9	8.4	6.6	15.0	3.6
Average	13.3	4.0	17.3	4.1

6. Forage production from sandy wastelands

There is. vast potential of augmenting fodder production from 6.36 million ha wasteland particularly sandy wastes and land with and without shrubs, which constitutes about 1/3rd of waste lands. Silvipastoral system with multipurpose tree species like *P. cineraria* and *Z. nummularia,* and grasses for increasing the forage production from degraded and waste lands should be a loud recommendation.

In arid zone of Rajasthan, sandy wastelands occupy 45% of the area. If stabilized and managed properly, these lands offer a huge potential productive site for afforestation and grassland development. Rehabilitation of sand dunes involves (i) protection against biotic influences, (ii) treatment of shifting sand dunes by fixing barriers (mulching-micro wind breaks) from the crests down to the heel of the dune, (iii) afforestation by direct sowing and or planting of suitable tree species,' and planting of seedlings or root slips of drought hardy perennial tussocky grasses. In such areas, grazing should not be allowed till the dunes are stabilized and grasses may be harvested manually to provide longevity to grass species.

7. Orans and Gochars as source of pasture grasses

Orans and gochars occupying about 62,000 ha, so called the Sacred forests, were traditional systems of grazing prevalent in arid western Rajasthan. These lands were attached to deities and were excellent source of biodiversity conservation. With passage of time there is a large scale encroachment on these grazing ground and these are on the path of degradation due to change in human values. By extra efforts, potential of such lands is required to be harnessed to supplement fodder during the stress.

8. Forage Production from Irrigated Areas

Forage production from a portion of cultivated area is practiced, where rearing of improved livestock like buffalo, high yielding cattle is common. Crop sequences for production of high yields of nutritious green fodder in irrigated areas are: Sorghum-berseem-pearl millet, maize+cowpea-berseeem-pearl millet and NB hybrid intercropped with berseem. Staggered planting of cropping sequences in rotational fashion by subdividing forage field in 2-3 plots, conservation of excess fodder as silage or hay during lush green period will solve many problems of drought affected region.

Recommended Fodder crops and their varieties

Sorghum	:	PC 6, PC 9, PC 23, HC 171, HC 260, SL 44, MP Chari, HC 136, Raj Chari 1, Raj Chari 2, Meethi Sudan, X 988, MFSH 3.
Pearl Millet	:	Giant Bajra, UUJ-IV-M, Rajko,
Maize	:	Africal Tall, Vijay Composite, Moti Composite.
Cowpea	:	EC 4216, Bundel Lobia 1, Bundel Lobia 2, UPC 5287

Berseem	:	Mescavi, Wardan, BL 22
Guar	:	FS 277, HFG 119, HFG 156, Bundel Guar 1, Bundel Guar 2
Hybrid napier	:	Pusa Giant, IGFRI 10
	:	Nandi, Kazungula, Narok
Guinea grass	:	Hamil, PGG 1, PGG 9
Dinanath grass	:	PS3, Bundel Dinanath 1, Bundel Dinanath 2

9. Judicious use of feed resources-Creation of feed/fodder banks

In large part of the country the dry and green fodder, straws and stovers are offered to animals without chaffing, resulting wastage of large portion of these resources. The practice of chaffing of fodder, straws and stovers should be popularized, which will enable more availability of energy for animal production. The technologies like urea treatment and complete feed by incorporating non-conventional feed resources, minerals, and complete feed blocks will also help in availability of more nutrients from existing quantity of feed resources.

Fodder banks: Storage of fodder from good to bad years was a traditional drought coping mechanism in arid region. One of the major issues that need greater attention is fodder bank creation in arid region, creation of fodder banks at Panchayat Samiti levels may be enforced for fodder security in famine years. The fodder bank concept should include bailing, densification, storage and transport of fodder to cater the need of deficient areas. The fodder bank could be the ultimate solution and contingent plan to meet the demand of lean period shortage particularly during drought and floods. These banks could be used for following feeds and fodder. Feed bank for ingredients unfit for human consumption, grasses from periphery of forest area, wastelands and farm lands stored as briquettes and high density stacks, crop residues in densified form, coarse cereals, legumes, haulms left after grain removal can also be stored.

10. Policy Intervention

Legislation

There is an imperative need for feed and fodder development policy at national level. Establishment for Feed and Fodder

Development Board can effectively work as a core instrument to achieve the policy goals for this sector.

To develop a grazing policy for the region, a policy for use of permanent pastures, as well as policy for animal migration is called for.

Strict enforcement of the policy decisions through dialogue and attractive schemes may help ease the pressure on degraded grazing lands.

Education

Forage production and research should be considered as a national agenda. Chairs should be created in Universities/SAUs. for teaching graduate and postgraduate students.

A sound syllabus be prepared, covering all major aspects of forage production and enforced at university level.

Conclusion

A wide gap in demand and availability of fodder and feed in arid western Rajasthan is reflected in low productivity of animal wealth, so important for sustainable livelihood particularly in drought years. There are opportunities for augmenting productivity of forages in arid regions. Traditional systems of "Orans" need to be rejuvenated and maintained to reduce pressure on grazing lands. Improved grassland technologies envisaging protection of area, reseeding with high perennial grasses, soil and water conservation, utilisation of pasture, management of excess forage and scientific methods of grazing can enhance productivity of these lands manifold. The non-conventional systems viz., silvipasture and horti-pasture, and forage production systems can be widely adopted on wastelands constituting 1/3rd of arid region in western Rajasthan.

The management techniques to ensure feed and fodder security are improvement in the production potential of fodder trees, bushes, range grasses and legumes for increased biomass availability.

Conservation of grasses germ plasm, breeding for biotic and abiotic stresses and introduction of short duration forage crops in the cropping system.

Appropriate strategies for efficient utilization of the existing feed resources and conservation of surplus fodder of normal years. To establish feed and fodder banks, storage and transport of fodder from excess to scarcity areas.

Policy intervention for promotion of fodder production, protection of grazing lands, minimize fragmentation of land holdings and use of tractors on grasslands.

Maximize forage production from non-conventional systems and problematic areas Training and education of farmers to popularize the grassland fodder production technologies.

REFERENCES

Balak Ram, Narain, P., Sharma, J. R., Nagaraja, R. and Jayanti, S. 2005. *Wasteland Management in Arid Western Rajasthan-* A joint publication of CAZRI, Jodhpur, RRSC, Jodhpur and MRSC, Hyderabad.

Dhabadgao, P.M. and Murwaha S.P. 1962. Productivity potential and palatability of grass spp. suitable for arid regions of India. *Ind. Jour. Agron.* 7: 1: 86-90.

Gajja, B.L. 2000. Economic evaluation of pasture management on marginal lands of arid zone of Rajasthan. In: *Training manual on Pasture Development and Fodder Production in Arid and Semi arid Areas.* Central Arid Zone Research Institute, Jodhpur, pp: 80-90.

Mishra, L.P. and Bhatt, R.K. 1996. Forage seed: research and technology. In: *Transfer* of *Forage Production Technology (Eds.* Shag Mal and Hagger, R.J.), pp: 91-99. Indian Grassland and Fodder Research Institute, Jhansi.

Narain, P. and Singh, R.S. 2002. Drought scenario in arid western Rajasthan. In: *Drought Management in Indian Arid Zone* (Eds. P. Narain, D.C. Joshi, S. Kathju and A. Kar). CAZRI, Jodhpur, pp. 16-29.

Narain, P. and Tiwari, J.C. 2005. Trees on agricultural fields: a unique basis of life support in *Thar* Desert. In: *Multipurpose Trees in the Tropics: Management and Improvement Strategies* (Eds. V.P.Tewari and R.L. Srivastava), Arid Forest Research Institute, Jodhpur.

Narain, P., Sharma, K.D., Rao, A.S., Singh, D.V., Mathur, B.K. and Ahuja, Usha Rani 2000. Strategy to combat drought and famine in the Indian Arid Zone. Central Arid Zone Research Institute, Jodhpur, 65 p.

Paroda, R.S., Mann, H.S. and Verma, E.M. 1980. Management of *Indian Arid Rangelands.* Technical Bulletin 4, Central Arid Zone Research Institute, Jodhpur.

Shankar,V. and Kumar, S. 1987 Grazing resources of semiarid and arid regions. In: Pasture and forage crops Research (Ed. Panjab Singh), IGFRI, Jhansi.440p.

Sharma, S.K. and Vashishtha, B.B. 1985. Evaluation of juju be-butte I grass hortipastoral system under arid environment. *Annals* of *Arid Zone,* 24: 303-309.

Singh, K.C. 1992. Grassland and pasture development in the arid regions of Rajasthan. In: *Rehabilitation* of *degraded arid* ecosystems (Eds. A.S. Kolarkar, D.C. Joshi and K.D. Sharma), Scientific Publishers, Jodhpur, pp: 121-126.

Singh, K.C. and S.D. Singh 1997. Water use and production potential of sewan *(Lasiurus sindicus* Henr). In: *Silvipastoral* systems *in arid and semi-arid* ecosystems (Eds. M.S. Yadav, Manjit Singh, S.K. Sharma, J.C. Tiwari and U. Burman). Central Arid Zone Research Institute, Jodhpur, 472 p.

Venkateswarlu, J., Anantharam, K., Purohit M.L., Khan, M.S., Bohra, H.C., Singh, K.C. and Mathur, B.K. 1992. Forage 2000 AD: *The Scenario for Arid Rajasthan.* CAZRI, Jodhpur; 32 p.

17

Shrub and Legume based Pastures for Sustainable Animal Production - Problems and Prospects

T.K. Bhati

Central Arid Zone Research Institute, Jodhpur - 342 003

Introduction

The hot arid tropical zone of India covers about 12 per cent of the country's geographical area, and occupies over 3.2 lakh sq. km of land located in parts of Rajasthan (61%), Gujarat (20%), Andhra Pradesh and Karnataka (10%), and Haryana and Punjab (9%). The human (23m) and livestock (20.5 m) population in this region is also very high. Even in normal rainfall year there is shortage of 36, 79 and 94 percent of dry forage, green fodder and concentrates respectively for the livestock and the animal feed shortage rises to 80 to 95% in drought year. To improve animal feed availability in the region besides fodder production from cultivated lands we have to improve the productivity of grazing lands (10 m ha) which is at present dismally low (0.35 t/ha). If not properly developed and scientifically utilized, these lands will continue to accentuate the process of desertification and soon may lead to serious environmental problems even for other parts of the country. CAZRI based on its over four decades of research has come out with full package of pasture establishment and utilization techniques for various sub-agro-climatic regions pertaining mainly to pasture grasses like *Cenchrus ciliaris, Cenchrus setigerus, Lasiurus sindicus, Dichanthium annulatum* and *Panicum antidotale* (Daulay and Bhati, 1989).

Fortunately nature has gifted these challenging areas with highly productive and useful species of pasture legumes and shrubs. There is great scope of optimal integration of legume components with pastoral grasses to increase the productivity, nutritive value, grand growth phase and palatability, of these pastures. Besides the pasture legumes can also fix the atmospheric nitrogen in the soil, thereby improving the productivity of grass component through biological N-fertilization. The response of pasture grasses to nitrogen application is well recognized (Bhati and Mathur, 1984) and proper integration of legume components with pasture grasses can help in the productivity enhancement of grasses through effective use of nitrogen fixed by the legume component.

Although at this Institute research work on adaptability and grazing behaviour of some of the pasture legumes viz. *Clitoria ternatea, Lablab purpureus, Indigofera* spp. has also been conducted. Besides several exotic pasture legumes viz., *Macroptilium atropurpureum* (Siratro), *Lotononis bainesii, Desmodium intortum, Centrosema pubescens, Stylosanthes hamata, S. scabra, S. viscosa, S. guianensis* were also studied for their adaptability in this tract. Except *Clitorea ternatea, L. purpaneus* and *M atropurpureum* non had shown much promise.

A. Experimental findings

Experimental findings of some of the studies carried out at this Institute (Bhati *et al.,* 1987) on various aspects of legume/shrub grasslands viz., germination, nodulation, compatibility, establishment and animal grazing are presented below for better comprehension of this important area of research.

Germination: Amongst pasture legumes the germination percentage of one year old seed of *Stylosanthus humilis* (16.0%) and *Indigofera cordifolia* (19.0%) were very poor as compared to *Atylosia scarabaeoides* (87.0%), *Lablab purpureus* (72%), *Macroptilium atropurpureum* (79%) and *Clitoria ternatea* (65.0%).

Nodulation: *L. purpureus, C. ternatea* and *M atropurpureum* nodulated very profusely (Table 1) in the field conditions. The per cent soluble sugar (2.09 and 3.62%), dry weight of nodules (122.15 and 38.67 mg/plant) were higher in *L. purpureus* and *C. ternatea* respectively, thereby, revealing their better nitrogen fixing ability. The results in general revealed that the pasture legumes viz. *L. purpurus, C. ternatea* and *M. atropurpureum* hold great premise for better establishment and biological nitrogen fixation in development of legume based pastures in arid regions.

Table 1. Abundance and chemical composition (%) of nodules in pasture legumes.

Pasture Legume	No. of effective nodules/ plant	Fresh wt. of nodules/ plant	Dry of wt. of nodules/ plant	Moisture %	Soluble sugar %	Nitrogen %	Starch %	Reducing sugar %
Atylosia scarabaeoides	9.84	19.23	16.48	16	3.4	36.54	1.28	0.05
Indigofera cordifolia	6.88	4.44	2.44	45.4	-	56.70	-	-
Lablab purpureus	29.2	423.00	122.15	71.92	2.09	41.79	0.27	0.07
Macroptilium atropurpureum	11.6	74.0	21.6	70.81	1.65	53.27	1.79	0.1

Compatibility: Investigations carried out to assess the feasibility of grass-legume association revealed that *L. purpureus* failed to regenerate while *Clitoria ternatea* continued to persist in association with perennial grasses viz. *L. sindicus* and *C. ciliaris* even during IIIrd and IVth years of establishment. Higher yields of forage were recorded in *L. sindicus* and *C. ternatea* (2:1) combinations.

Soil and Water Conservation: Protection from biotic factors and adoption of improved soil and water conservation measures (gully control, staggered contour trenches, vegetative barriers etc.) could result increase in the density of several naturally occurring shrubs in rocky degraded grassland substantially in a period of 5 years (Table 2). The increase in density was highest (872%) in *Ziziphus nummularia* and was least in *Grewia tenax* (187.5%). All these shrubs are very important for improvement in nutritive value of the fodder produced on one hand and enhancement of grazing period by 1 to 2 months on the other.

Table 2. Increase in the density of natural woody perennials under protection and soil and water conservation measures after 5 years.

Woody perennial	Density (No. of plants/ 30 ha)		
	Initial	Final	% Increase
Capparis decidua	103	315	205.8
Maytenus emarginata	41	221	439.0
Acacia jacquemontii	10	76	650.0

L ycium barbarum	36	193	436.1
Grewia tenax	8	23	187.5
Ziziphus nummularia	44	438	872.8
Total	266	1458	448.12

Table 3. Forage (t/ha) and seed (q/ha) productivity of legume/shrub pastures in establishment years

Type of pasture	Forage yield (t/ha)					Seed yield (q/ha)				
	1982	1983	1984		Total	1982	1983	1984		Total
	Grass	Grass	Grass	Legume/ Shrub		Grass	Grass	Grass	Legume / Shrub	
(A) *C. ciliaris*	1.40	1.40	2.78	-	2.78	1.10	1.62	2.76	-	2.76
(B) *C. ciliaris* + *C. ternatea* (1:1)	1.31	1.42	1.14	1.07	2.21	0.95	0.96	0.96	1.82	2.88
(C) *C. ciliaris* + *C. ternatea* (2:1)	1.42	1.31	1.49	0.46	1.95	1.04	1.32	1.42	1.34	2.76
(D) *C. ciliaris* + *Z. nummularia* (1:1)	1.00	0.95	0.83	0.17	1.00	0.60	0.85	0.93	1.50	2.43
(E) *C. ciliaris* + *Z. nummularia* (2.1)	1.20	0.90	1.62	0.17	1.79	0.80	1.45	1.18	0.89	2.07

Legume-shrub grasslands

Preliminary studies on the role of shrubs in pasture lands was conducted by Kaul and Ganguli (1963) and they observed highest production (1.0 t/ha) on the natural grasslands of which 14% area was covered with *Z. nummularia* shrub. Sharma and Vashistha (1980) also reported highest forage yield (1.2 t/ha) in *C. ciliaris* + *Z. nummularia* (210 plant/ha) pasture. Research programme for evaluation of shrub/legume based pastures under animal grazing system (mixed flock of sheep and goat) at Jodhpur (av. rainfall 357.6 mm) with *C. tematea* as a pasture legume and *Z. nummularia* as a shrub in varying proportions with *C. ciliaris* [A. 1:0 (pure grass), B. 1:1 (grass + legume), C. 2:1 (grass + legume), D. 1:1 (grass + shrub), E. 2: 1 (grass + shrub)] on a paddock of 0.75 ha size for each treatment revealed that the pasture No. B to E besides giving acceptable levels of grass yield (Table 3) also produced highly nutritive fodder for the animals from the legume/shrub component (Bhati *et al.,* 1987). However, introduction of grazing resulted heavy pressure specially on legume component and resulted in complete mortality of *C. ternatea* in pasture No. B and C. As such it appeared that the pasture legume although good for stall feeding may perhaps not be able to sustain the animal grazing pressure in arid zone conditions. Contrary to legume, *Z. nummularia* could sustain the heavy browsing pressure of animals well and no mortality was observed. The live weight gain and wool hair production was higher in the pastures having these components (Table 4).

Since all the three years (1984, 1985, 1986) have been the drought years the animal grazing could not be continued on year long basis and in lean periods animals were fed from outside fodder resources. It is therefore suggested that we should develop a three tier system involving top feed tree species, shrubs/legumes and grass component in suitable proportion for year long grazing and sustainability of animal production from these lands. In fact it can be an important area of pasture research programme.

Table 4. Live weight gain (kg/ha) and wool/hair (kg/ha) production from sheep and goat for legume/shrub pastures

Pastures	Live weight gain (kg/ha)						Wool/hair production (kg/ha)			
	Ram lambs			He-goat			Ram lambs		He-goat	
	Dec. 84 to June 85	Oct. 85 to Dec. 85	Sept. 86 to Jan.87	Dec. 84 to June 85	Oct. 85 to Dec. 85	Sept. 86 to Jan. 87	Dec. 84 to June 85	Oct. 85 to Dec. 85	Dec. 84 to June 85	Oct. 85 to Dec. 85
(A) *C. ciliaris*	18.5	3.0	3.0	22.0	4.5	10.0	4.5	2.9	0.85	0.95
(B) *C. ciliaris* *C. ternatea* (1:1)	19.0	-	-	2.0	-	-	5.0	-	1.25	-
(C) *C. ciliaris* *C. ternatea* (2:1)	19.5	-	-	22.0	-	-	4.1	-	1.20	-
(D) *C. ciliaris* *Z. nummularia* (1:1)	25.0	13.0	9.0	23.5	16.0	11.5	4.2	3.2	1.3	1.05
(E) *C. ciliaris* + *Z. nummularia* (2.1)	24.5	16.0	16.0	24.0	170	16.0	4.75	3.1	1.35	1.25

B. Economic Evaluation

(i) Establishment Cost: Establishment of shrub based pastures was found costlier than the pure grass pasture because of the added cost of shrub establishment The total expenditure towards the pasture establishment was Rs. 2187.00, 2275.00, 2058.50, 4608.00 and 4241.50 for pasture numbers A, B, C, D and E respectively (Bhati *et al.*, 1987). Abnormal rainfall pattern in the establishment phase (1982-83, 1983-84) also added to the cost particularly in *Ziziphus* based pastures.

Table 5. Year wise grass monetary returns (Rs./ha) from legume/ shrub grasslands.

Pasture	1982-83	1983-84	1984-85	1985-86	1986-87	Total over 5 years
(A) *C. ciliaris*	832.50	3520.50	1139.30	312.75	412.50	62177.55
(B) *C. ciliaris* + *C. ternatea* (1:1)	456.00	930.30	1125.00	-	-	2511.30
(C) *C. ciliaris* + *C. ternatea*	241.50	1919.40	1125.10	-	-	3286.00
(D) *C. ciliaris* + *Z. nummularia* (1:1)	51.75	1425.43	1295.40	881.50	600.00	4254.08
C. ciliaris + *Z. nummularia* (2:1)	238.50	1652.80	1309.55	994.45	600.00	4795.3

(ii) Gross monetary returns: The monetary returns from these pastures were estimated on the basis of forage, seed, green leaf *('Pala')*, bushwood (fire wood), manure, wool/hair and live weight gain. Although based on 5 years of estimations pure pasture of buffel grass gave higher returns (Rs. 6217.55/ha) but the returns were also satisfactory from shrub-based pastures i.e. D (Rs. 4254.08 ha) and E (Rs. 4795.30/ha). Again the yield of shrub component is also bound to increase over time. Hence these pastures are expected to give better returns in long run. Further as reflected in Table 4. The live weight gain and wool production were also conspicuously higher in these pastures and this trend will continue to turn them more and more profitable with the passage of time. The year wise monetary returns from these pastures are given in Table 5. In 1985 and 86 the returns were calculated only on the basis of animal productivity and it clearly reflected the superiority of these pastures over pure grass pasture. Since the prices of wool and meat were low during these years i.e. Rs. 25/kg and 22/kg respectively as reflected in low monetary returns over previous years where higher seed price (Rs 25/kg for grass, Rs. 5/kg for legume) reflected in high returns.

Conclusions

From the foregoing discussion it is evident that nature has bestowed arid regions not only with highly productive and useful species of pasture grasses but also of legumes and shrubs. If properly dovetailed with pasture grasses these legumes/shrubs can help in 1 improvement in pasture productivity and quality, 2. enhanced palatability and grazing period, 3. improved soil fertility and productivity, 4. efficient soil and water conservation and ultimately; 5. socio-economic stability through enhanced livestock productivity and quality. Although *Z. nummularia* has been identified as suitable shrub in association with pasture grass *C. ciliaris* (1:2), but there are many indigenous *(Calligonnum polygonides, A. jacqumantii, Mimosa hamata, Dichrostachys cinerea* etc.) and exotic *(Dichrostachys nutans, Colophospermum mopane* etc.) shrubs which need to be evaluated for their suitability in development of shrub based pasture systems befitting different agro-climatic situations in the desert. Similarly there are several pasture legumes *Rhyncosia minima, A. scarabeoides, Indigofera* spp. etc.) which need to be investigated for their compatibility and proportion with pastoral grasses both under stall feeding and grazing management at can be concluded that active people's participation is of paramount importance in development and management of these systems on community/private lands.

REFERENCES

Bhati, G.N, Bhati, T.K. and Shankarnarayan, K.A. 1987. Economics of pasture establishment and evaluation in arid regions. *Annals Arid Zone* 26(4): 241-247.

Bhati, T.K. 1984. Grass component in silvi-pastoral system with special reference to Indian Arid Zone *Agroforestry in arid and semi arid zones* (Ed. Shankarnarayan *et al.*, 1984), CAZRI Publication No. 24 150-155.

Bhati, T.K. and Mathur, B.K. 1984 Effect of nitrogen, phosphorus and farmyard manure on forage yield and nutritive value of *Cenchrus ciliaris* Linn. *Indian Journal Agronomy* 26(1) 80-86.

Bhati, T.K., Shankarnarayan, K.A. Bhati, G.N. and Mertia, R.S. 1982-86. Performance of sheep and goat on legume/shrub grasslands of western Rajasthan. *CAZRI Annual Progress Report* 1982-86.

Daulay, H.S. and Bhati, T.K. 1989. Crops and pasture production potential of sandy soils. In the book on *Review of Research on Sandy Soils*, CAZRI, Jodhpur, pp. 245-262.

Kaul, R.N. and Ganguli, B.N. 1963. Fodder potential of *Ziziphus* in the shrub grazing lands of arid zones. *Indian Forester* 89: 623-630.

Sharma, S.K. and Vashishtha, B.B. 1980. Horti-pastoral system in arid zone. *Annual Progress Report,* CAZR1, Jodhpur.

18

Management of shelter for amelioration of environment stress for optimum production

A.K. Patel

Centeral Arid Zone Research Institute, Jodhpur - 342 003

Introduction

A major part of arid region is characterized by harsh environmental conditions due to low and erratic rainfall, poor quality of soil, higher variation in air temperature, high wind velocity, high evopotranspiration etc. Such areas are suitable mainly for silvi-pastures and livestock rearing. Animals of this region are mostly maintained on extensive grazing on rangelands, where these animals are not provided any kind of shelter during grazing/browsing, but in extremes of summer animal seek shelter under the tree shade. Whereas, during extremely cold conditions at night in winters the animals take protection by the sand dune, dense trees or shrubs. Extremes in the environmental conditions can have enormous effects on physiology and productivity of farm animals (Singh *et al.,* 2001 and Singh and Upadhyay, 2009). Numerous studies indicate that the provision of proper shelter improves livestock performance through moderation of environmental factors (Patel *et al.,* 2001, Singh *et al.,* 2003 and Patel *et al.,* 2007). A long period of summer, monsoon and post monsoon are stressful seasons in arid and semi-arid regions. Animals should be protected from direct solar raditation during the hottest hours of the day to ameliorate the effect of heat stress (Shinde *et al.,* 2002).

(i) Cattle adaptability

Livestock are homeotherms, i.e., they must keep their body temperature within a moderately narrow range to work efficiently. Tharparkar and Rathi breeds of cattle are well adapted to arid environmental stresses. The heat tolerance co-efficient values of free grazing animals of these breeds were 88.88 ±1.92 and 89.87 ± 2.51, respectively. During summer season, there was an increase in pulse rate and respiration rate, these parameters were found to be the highest in June and lowest in January. The daily water consumption per 100 kg liveweight was also higher (13.27 L) during May and lowest (5.62 L) during December (Anon., 1990). Physiological norms of Rathi cattle at weekly interval was analysed. The highest 39.7°C and lowest 37.1°C rectal temperature was recorded in June afternoon and January morning, respectively. The diurnal variations were observed in rectal temperature, pulse rate and respiration rate, the magnitude of these traits was higher from April to September (Anon., 1994). Rathi cattle had significantly higher percent packed cell volume, haemoglobin and total body water in comparison to Tharparkar cattle, suggesting better adaptability of Rathi cattle in arid environment of Bikaner region.

(ii) Adaptability of small ruminants

Water is the most limiting factor for survival in desert. The long term effect of heat and water stress on sheep physiology was studied at CAZRI. These studies observed unusual ability of the desert breed to maintain circulation even when faced with considerable haemoconcentration. When water was reduced to below 75% of normal daily requirement, there was a steady decline in the body water stores. On an average, there was an 18% loss in body weight after remaining without water for 3 days during winter and 25% during summer. Interestingly, the digestibility of crude fibres was found to be increased in water restricted sheep while nitrogen balance was not affected due to water stress. The rate of passage of feed is slow in water restricted sheep in comparison to normally hydrated animals.

Responses of sheep to solar exposure: Hogget of native Nali, Chokla and their crosses with Rambouillet and Soviet Merino were maintained in shade or sun during summer. After first recording, the animals of sun and shade groups were exchanged and the observations were continued. The respiration rate, rectal temperature, skin temperature and fleece temperature 1 and 2 cm above the skin were higher in the solar exposed group as compared to the animals maintained in shed and higher at 16 and 13 h than at 9 h. In general the rectal temperature of the crossbred animals was lower than the native; lower in Merino x Nali and highest in Nali. It was observed that the animals maintained under hot sun experienced greater stress as compared to the animals under shade. It also appeared that in crossbreds the thermolytic mechanism was activated at a lower ambient temperature resulting in their higher respiration rate and lower rectal temperature (Kaushish *et al.*, 1989). The physiological response viz. respiration rate and heart rate were lower in sheep provided shade and nutritionally restricted or fed ad lib in comparison to those which were exposed in sun (Hooda and Naqvi, 1990).

Physiological responses of sheep and goats to exercise stress under arid conditions: The water intake was higher in exercise group of sheep and goats during summer and winter season. The body weights of both could not be maintained during summer season. Significant diurnal differences were observed in rectal temperature, pulse rate and respiration rate. The rise in rectal temperature was more in the afternoon in exercise group. The differences in rectal temperature, pulse rate and respiration rate between summer and winter seasons were significant. The rectal temperature was lower in goats than in sheep. Differences in rectal temperature between groups were non-significant. It can be concluded that under arid conditions of Rajasthan, Magra sheep and Marwari goats can safely be grazed within 8 km from their watering point without any ill effect on their health when the average atmospheric temperature is around 44°C (Kaushish and Mittal, 1994).

Effect of water restriction and thermal stress on physiological norms: The goat breeds of Beetal and Black Bengal were divided into 2 groups, in one group water was not withheld, whereas it was withheld for 36 h in other group before exposure to thermal stress (43° ± 0.5°C, RH 55%). Increase in all the 3 body responses was seen under thermal stress. Maximum respiration rates were observed when goats attained rectal temperature between 40.3°C and 40.5°C. The rectal temperature level at which maximum respiratory frequency was observed, was lower when the animals were under combined stress of thermal exposure and water restriction. Significant differences were observed between breeds and groups in pulse and respiration rates (Kaushish *et al.*, 1987).

Effect of heat and water stress on cortisol level in goat: This studies was conducted on 5 to 6 months old male kids (9 Beetal and 9 Black Bengal), which were exposed to 43.0 ± 0.5°C for 2 h. The blood hormonal estimation was done in above mentioned breeds. Cortisol concentration in control, thermal stress and thermal stress coupled with water stress groups were 17.6± 2.68, 54.3±1.71 and 38.5±5.13 ng/ml in Beetal and 48.1±3.60, 43.7±5.32 and 55.5±5.72 ng/ml in Black Bengal goats, respectively, Significant increase in cortisol level was observed in Beetal on exposure to thermal stress, but the increase was non-significant in Black Bengal. It is conducted that Black Bengal were more adapted to hot humid conditions than Beetal (Kaushish *et al.*, 1997b).

(iii) Effect on milk production performance

The milk yield of all mammalian species undergoes seasonal variation. The optimum environmental temperature for lactation depends on the species, breed and degree of tolerance to heat or cold. The milk yield of Holstein cattle declines at temperature above 21°C, in case of Brown Swiss and Jersey cattle it declines at about 24° to 27°C whereas milk yield of Zebu cattle declines only above 34°C. Observations in India on cross-bred cows has shown that the productivity of cross-bred of exotic with indigenous breeds were not seriously affected by the hot dry climate in north western parts. On the other hand in hot humid coastal and eastern region of the country with heavy rainfall, the productivity was much lower. Shinde *et al* (1990) observed that milk yield per day decreased with simultaneous increase in ambient temperature and humidity. Dutta *et al* (1995) reported that temperature and humidity together explained 62.75% of variation in milk yield/cow/day, which was higher than that explained by either temperature or humidity alone. In a study of thermal stress and milk production in half-bred Holstein crosses (Karan Fries) and zebu cattle Sahiwal, the Karan Fries showed higher sensitivity for temperature rise (Singh and Upadhyay, 2009). When THI crossed the index value of 72 it impacted milk production negatively, the average milk production of Karan Fries and Sahiwal animals was 13.4 and 6.6 litre/day at around THI of 72. The respective milk production was decreased to around 9 and 5 litre/day in Karan Fries and Sahiwal cows at around THI of 82. The increase in each unit THI impacted in reducing the milk production by 0.43 and 0.16 litre/day/animal, respectively in the range of 72-82 THI. The effect of summer season on The productivity of Tharparkar cattle was not significantly influenced by season of calving in arid zone though the 305-days lactation was higher in winter season calvers (1438.8 litre) than summer season (1274.9 litre), however, the total lactation yield was more or less similar in both season due to one month longer lactation length in summer season calvers (Patel *et al.*, 2000). But the milk fat content of Tharparkar cattle was found to be lowered in both hot dry (4.30%) and hot humid season (4.64%) in arid zone whereas, higher percentage of fat was observed during Autumn (5.24%) and winter season (5.25%) (Patel *et al.* 1994). Though the Marwari goat breed of Thar desert is well adapted to desert vagaries, the productivity of these animals is affected adversely due to extremes of climatic conditions. Patel *et al.* (2001) observed lower lactation yield in Marwari goats during winter when housed in thatched roof animal shelter in comparison to closed type pucca animal shelter in arid region of north-west part of India. These goats yielded 38 lit more milk in a lactation period of 154 days with higher milk fat 4.77%. In a another study they found 22% more lactation yield in Marwari goats when these were housed in closed type Improved Animal Shelter than open type Tradition Animal Shelter during winter season (Patel *et al.,* 2007). The reason for higher milk yield in Marwari goats under improved Animal Shelter might be due to protection of goats against cold weather in winter season.

(iv) Effect on reproductive efficiency

Heat stress affects the reproductive physiology in farm animal. In a study of effect of thermal stress on superovulatory response, the occurrence of estrus was late (30.6+1.2 h) with shorter estrus period (31.7+3.6 h) in a group of ewes exposed to thermal stress in comparison to group of ewes which were maintained under shed (Naqvi *et al.,* 2004). However, the thermal protected ewes exhibited estrus earlier (25.5+1.1 h) and the estrus period was longer (37.7+1.6 h). The reproductive performance of Tharparkar cattle in terms of dry period and calving interval was not significantly influenced by season of calving but the results of the study showed that comparative lower dry period (101.9+23.3 days) and calving interval (414.5 + 27.9 days) was observed in winter season calvers than summer and rainy season calvers (Patel *et al.,* 2000). On the other hand it was also observed that buffaloes calved in rainy season had higher reproductive efficiency in terms of lower service period (198.4+15.1 days) and calving interval (501.4 + 15.3 days) as reported by Patel and Tripathi, 1998. The optimum service period was found 2 months in Tharparkar cattle where the milk production and reproduction efficiency was higher under arid conditions (Patel *et al.,* 2001).

The reproductive performance of Kuchhi, Marwari and Beetal breeds of goats was studied. The average age at puberty, first conception and first kidding ranged from 311.2±5.43 to 560.0±6.52, 387.4±3.75 to 605.9±4.59 and 509.4 ± 3.45 to 760.3±4.50 days, respectively. The average values for service period and kidding interval were shortest 157.00 ± 1.30 and 305±4.56 days, respectively in Kuchhi. The post-partum interval and getting interval were shortest in Kuchhi and longest in Beetal. The tupping and concentration were higher in Kuchhi (Kaushish *et al.*, 1994).

(v) Effect on growth performance

The body weight gain in growing age is an important aspect for any domestic animal reared for economic benefit. The climatic stress particularly due to heat during growth period of animal reduced the body weight gain. Higher body weight gain was observed in the 6 month old male kids of Marwari breed, when reared under closed type Improved Animal Shelter during summer months. Kids gained 9.52 kg body weight in Improved Animals Shelter than 7.52 kg in Traditional Animal Shelter during 6 months period of experiment from May to October (Patel *et al.,* 2007). The month of birth of lambs affects the body weights, lambs born in January to February were heaviest at 6 months body weight (30.6+0.93 kg) with decreasing trend in lambs born in subsequent months i.e. March to December time period (Poonia, 2004). Supplementation for higher growth in goats is essential, since the availability of forage is scanty in grazing lands of arid zone. In an experiment of supplementation feeding in four breeds of goats in arid zone, the higher body weight gain was observed in Jhakrana (54%), followed by Jamunapari (53%), Parbatsari (47%) and lowest in Marwari (16.2%) (Patel *et al.,* 1998). Rohilla and Khem Chand (2004) also observed a significant growth in Marwari kids when they were supplemented with concentrate feeds at daily and alternate day. The body weight gain in the end of experiment was higher in both supplemented groups 52.23 and 47.7 g/d, respectively in comparison to control group (39.9 g/d).

Improved Animal Shelter with east-west orientation

Provision of plenty of sunlight inside of animal shelter

Use of thatched panels to reduce the heat load from roof

Proper feeding space for animals in side shelter

Fig. 1

(vi) Present status of Shelter management practices in the arid region

The survey study conducted in arid region revealed that 80% farmers provide thatched shelter to cattle and only 10% farmers provide thatched shelter to small ruminants. Most of the animal houses were having rectangular shape but few of them were in round shape. Almost all animal houses were having soil floor, and pucca floor was only found in commercial production system near city areas. The local materials which were used to construct the animal shed were basically non edible parts of grasses, shrubs and trees etc., which are freely available in the villages. Feed manger was found in almost all animal shelter, but the water trough was found only 20% in animal houses because most of the animals in the villages drink water from a common water body. No proper attention for animal shelter was given, low cost or free local materials were only used to construct the animal houses. Farmers were found more concerned for animal housing during winter season than other seasons. Three types of shelters for livestock were observed in the rural areas of the arid zone, viz; open housing system (Barbed enclosure without any roof), Kutcha (Thatched roof house) and Pucca housing system which was a closed type pucca house (Patel *et al.*, 2005).

Study conducted at institute farm showed that pucca house was comfortable in comparison to the other two types of traditional animal shelters (Patel *et al.*, 2001). Based on the study of different factors like orientation, wind direction, sun movement, air temperature, local materials for construction etc. (Fig. 1). Improved animal shelter consists of angle iron frame, asbestos sheet and stone slabs. Heavy duty angle iron (50 x 50 x 6 mm) frame was fabricated and fixed on the ground by grouting. Seven heavy duty trusses were also fabricated and fixed at ten feet interval for proper distribution of asbestos roof load.

The orientation of the shelter was kept east-west direction; the long axis of shelter is 60 ft which is open from south direction and north direction. Hence, this orientation with generous provision for ventilation/air movement to help dry up threshold will be most suitable. Furthermore, the south facing shelter has the advantage of receiving larger solar radiation during winters than during summer. Small over hang on south faced can cut off direct solar radiation during summer and allow it during winter. It also gives protection from rains. The open space of north axis was covered by plastic curtain to minimise the entry of cold winds inside the shelter during winter.

The double slope roof of the shelter was made by corrugated asbestos cement sheets (1 x 3 mtrs.) on angle iron frame. The roof height is 10 feet from centre and 7 feet from sides. 4 inch thick thatch panels were fixed just below the asbestos sheets to provide insulation. The side walls were erected using stone slabs with 4ft height in north faced and 3 ft. in south faced. The floor of this shelter is elevated 1 ft. from the ground and filled with calcified soil to give compactness, which also absorbed the urine and faecal material of the goats.

The feeding manger of 50 ft. length was constructed along with the wall of north side of shelter by using stone slabs of this side. The height of manger was kept 1 ft. 9 inch from the ground level with 9 inch depth and 12 inch width for holding ample feed material. This manger is also facilitated with slanting slope from outer to inner side which helps in sliding of feeding materials while feeding to the goats. A ladder type structure of 1" diameter G.I. Pipe has also been fixed on outside wall of the manger to prevent entry of the goats inside of the manger to avoid the wastage of feed materials. In the same direction a 10 ft. long water trough with similar dimensions was also constructed in the corner of the shelter.

Microclimate of Improved Animal Shelter

Higher minimum temperature was recorded in Improved Animal Shelter (IAS) during winter from October to March ranging from 12.7 to 21.6°C. Higher value of minimum temperature for summer months was also recorded in Improved Animal Shelter from April to September, which ranged from 23.3°C in April to 28.20°C in June. Whereas, comparative lower values of minimum temperature were recorded in Traditional Animal Shelter (TAS) round the year. The difference of minimum temperature between IAS and TAS was more (1.42°C) during winter and less in (0.35°C) in summer months (Mishra *et al.*, 2006). The maximum temperature in IAS was found lower during summer months from March to September in comparison to TAS, while only in the month of July maximum temperature was almost equal in both animal shelters may be due to rainy season and higher level of moisture in air. However, during winter months i.e. November to February, the values of maximum temperature in IAS were found higher than TAS.

In IAS, the higher value of minimum temperature was mainly due to the absence of direct entry of air temperature inside close type animal house, secondly the minimum temperature which was attained usually in absence of thermal radiation during the time beyond midnight, remained higher in closed type animal shelter, because of inadequacies in loss of absorbed heat and low ventilation also add to this effect. The lower maximum temperature in IAS during summer months, could be due to the failure of direct sunlight to reach inside closed type animal shelter, which prevented the rise of ambient temperature much beyond that in open, whereas the refection radiation in addition to direct radiation caused higher maximum temperature in TAS. The higher level of relative humidity in IAS has been observed, which could be due to low ventilation means and less air flow inside shelter.

Conclusions

Although, the livestock breeds of arid region are resistant to high and low environmental temperature, but they do suffer from extremes of hot and cold seasons, when exposed to upper and below critical temperature. Tree shade has been found to be common available shelter to the livestock, with its limitations. Closed type animal house with open space for free movement for livestock, has been found to be most suitable animal shelter system in arid zone.

REFERENCES

Anonymous (1990). Responses of Rathi cattle fed on *Prosopis cineraria* (Khejri) leaves. Annual report, CAZRI, 1989-90, 65 p.

Anonymous (1994). Performance of Rathi cows and calves under arid conditions Annual Report, CAZRI 1993-94, pp.90-91.

Dutt, T., Taneja, V.K. and Singh, A. (1995). Effect of climatic variables on average daily milk production in crossbred cattle. *Indian Journal of Animal Sciences*, 65(a) : 1004-1007.

Hooda, O.K. and Nagvi, S.M.K., (1990). Effect of thermal load and feed restriction on the relative adaptability of Malpura and Avikalin sheep in semi-arid region. *Indian Journal of Animal Sciences*, 60(5):608-611.

Kaushish, S. K and Mittal, J.P. (1994). Physiological responses of sheep and goats to exercise stress under arid conditions. *Indian Journal of Animal Sciences* 64:423-424.

Kaushish, S.K., Georgia, G.C. and Sengupta, B.P. (1987). Effect of heat and water restriction on physiological response of Beetal and black Banal goats. *Indian Veterinary Journal* 66:852-856.

Kaushish, S.K., Dharam Singh and Satya Paul (1994). productive and reproductive performance of different breeds of goats under semi-arid conditions. *Indian Veterinary Journal 71* : 565-568.

Kaushish, S.K., Sengupta, B.P. and Georgie, G.C. (1997). Effect of thermal stress and water restriction on cortisol level of Beetal and Black Bengal goats. *Indian Journal of Animal Sciences* 63:1277-1279.

Mishra, Dinesh, A.K. Patel and Harpal Singh (2006). Development of Improved Animal Shelter for Arid Region. National Symposium on "Livelihood Security and Diversified Farming Systems in Arid Region". January 14-16, 2006, CAZRI, Jodhpur. PP:59-60

Naqvi, S.M.K., Maurya, V.P., Gulpani, R., Joshi, A. and J.P. Mittal (2004). The effect of thermal stress on supervolatory response and embryo production in Bharat Merio ewes. Small Ruminant Research 55:(1-3) pp. 57-63.

Patel, A.K. H.C. Bohra and. S.K. Kaushish (1998). Effect of supplementation on comparative weight gain of certain goat breeds in arid zone. *Indian Journal of Small Ruminants* 4 (2), 91-93.

Patel, A. K., Rohilla, P.P. and A. C. Mathur (2007). Performance of Marwari kids and goats under improved and traditional animal shelter. Indian Veterinary Journal 84:1069-1071.

Patel, A.K. and Tripathi, V.N. (1998) Effect of non genetic factors an economic traits of Surti buffalo. *Indian Journal of Animal Sciences,* 68(6): 566-569.

Patel, A.K., A.C. Mathur and Dinesh Mishra (2005). Traditional shelter management practices for livestock in arid region. National Conference on "Bridging Gap Between Ancient Technologies to Increase Agricultural productivity". December 16-18, 2005 CAZRI, Jodhpur. PP: 53-54

Patel, A.K., Mathur, A.C. and MIttal, J.P. (1994) Factors affecting composition of milk in Tharparkar cows maintained under arid conditions. Indian Journal Animal Production & Management, 10: 102-105.

Patel, A.K., Mathur, B.K. and Kaushish, S.K. (2000).Effect of shelter management on body weight and milk yield of Marwari goats under arid condition. Int. Conference on Small Holder Livestock Production Systems in Developing Countries, Nov. 24-27, 2000, Thrissur, Kerala.

Patel, A.K., Paharia, Shailesh and Nagpaul, P.K. (2001). Productive performance of Marwari goats under different shelters in arid region. Indian Journal Animal Production and Management 17: 76-78.

Poonia, J.S. (2004). Growth performance of Munjal lambs. *Indian Journal of Small Ruminants* 10 (2), 137-139.

Rohilla, P.P. and Chand, K. (2004). Effect of supplemental feeding on growth of kids and milk yield of Marwari goats. *Indian Journal of Small Ruminants* 10 (2), 143-146.

Shinde, A.K., Bhatta, Raghavendra., Sankhyan, S.K. and Verma, D.L. (2002). Effect of season on thermoregulatory responses and energy expenditure of goats on semi-arid range in India. The Journal of Agricultural Science Cambridge University Press 139:1:87-93

Shinde, S., Taneja, V.K. and Singh. (1990). Association of climatic variables and production and reproductive trait in crossbred. *Indian Journal of Animal Sciences,* 60(1):81-85.

Singh, M., Tyagi, A. K., Mehla, R. K. and Singh, C. (2003). Milk production performance of cows under two different sets of management during summer. Indian Journal of Animal Sciences 73(6): 692-694.

Singh, P., Dixit, V.P., Singh, B. and Georgie, G.C. (2001). Plasma aldosterone and major circulatory and urinary electrolytes in buffalo calves as affected by peak summer and winter stress. International Journal of Animal Science 16(1):161-166.

Singh, S. V. and Upadhyay, R. C. (2009). Thermal stress on physiological functions, thermal balance and milk production in Karan fries and sahiwal cows. The Indian Veterinary Journal 86:141-144.

19

Management and Feeding of Sheep During Limited Feed Resources

B.S. Rajpurohit

Central Sheep Breeding Farm, Hisar - 125 001

Introduction

In many ways cultural life in India is closely associated with livestock. The contribution of livestock to the national economy is indeed vast. The livestock production systems generally followed in the country based upon mixed farming in which crop-livestock integration is the tradition. Cattle and buffalo are raised in small mixed farms. Sheep and goat raising through small and marginal farmers and agricultural labourers offers good scope in the country. Sheep and goat contribute about half of the total meat production of the country. As per 2003 census the sheep population with 61.47 million sheep India ranks 4th in the world (4.5% of world sheep population). Sheep producing about 50.94 million kg (2002-03) wool with average of 800 grams/sheep and contributes about 170 million kg of mutton and 40 million kg of skin.

Nutrition plays a major role in the overall productivity, health, and well-being of the sheep flock. Because feed costs account for approximately two-thirds of the total cost of production, it is important that producers consider nutrition management a top priority. Nutrient requirements of sheep vary with differences in age, body weight, and stage of production. During the grazing season, sheep are able to meet their nutrient requirements from pasture and a salt and mineral supplement.

Forages constitute 75 to 90% of the total diet for sheep. Sheep are excellent converters of forage to meat and fiber and consume a wide variety of forages, and selectively graze numerous weeds and other pasture menaces. Companion grazing of sheep with other species of livestock, such as cattle or goats, results in greater pasture utilization and higher quality pastures than when a single species is grazed alone.

Pasture

Permanent pasture should be the predominant source of nutrition for the sheep flock. Intensive sheep production systems where the sheep are housed and fed harvested feeds are not as profitable as more extensive production systems where they harvest their own feed. When a sufficient quantity of forage is available, sheep are able to meet their nutrient requirements from forage alone along with a supplemental source of salt and minerals.

Drought action plan

Since droughts are part of life for sheep farmers in the country, Planning and decision making must be done as soon as you recognize the possibility that the poor season may progress to a drought. If you leave the decisions until the drought worsens, many of the management options available early may be closed to you. Prices for sale stock usually drop dramatically, pastures dries up, fodder prices generally soar and off farm employment becomes difficult to find (Narain *et al.*, 2000).

Plate 1. Concentrate feeding to lambs

However, each drought brings its own set of difficulties. How well you survive the drought will depend on the initial plan of action and the modifications undertaken to the strategy as the drought progresses. The first step is to list the farm's financial and physical resources so that the effects of various strategies, both short and long term, can be calculated. Water is probably the first thing to consider as if this resource is inadequate, then it will be difficult to retain large numbers of stock. The next step in choosing a drought action plan is to estimate when you think the drought will break. This will affect your calculations on how long you will be feeding sheep, how much it will cost and whether you will decide to sell stock or not. It is best to over estimate the time you expect to hand feed your sheep to be on the safe side.

Migration

Sending sheep away on migration is sometimes more economical than feeding and the time saved from feeding might be more usefully employed. It also releases more feed for the stock remaining on the property. Don't forget to inquire in your own district, especially early in the drought, as you may be able to find ungrazed paddocks for rent. However, if the drought becomes more widespread, it is harder to find place nearby for grazing the animals and the cost rises rapidly. It may then be cheaper to feed stock at home. It may also be costly and impractical to supervise sheep (especially lambing ewes) at a distance.

Selling

Selling sheep is one option where early planning and action is advantageous. In particular, selling decisions should be made before stock have lost too much condition to be saleable and market prices have started to drop.

Feeding

The class and stage of production of the animals dictate the type and quality of forage to be grazed. Lactating ewes with lambs are placed on the highest quality pasture available to promote desired levels of milk production and lamb growth. Dry, non-pregnant ewes or ewes in early to mid-gestation are placed on lower quality forages or serve as second grazers behind young, growing lambs. Strategies that match stage of animal production with type and quality of forage improves overall forage utilization while maintaining optimum animal performance.

The aim of feeding sheep in a drought is to maintain weight in dry sheep and to meet the requirements of late pregnant and milking ewes. You should allow for enough growth of lambs and weaners so that they do not suffer permanent checks. Feeding should be started well before the sheep become weak. It may take some time before they become accustomed to hand feeding and begin eating their ration. If sheep have lost too much condition before feeding has begun, or before they readily accept grain, it may be hard to lift their live weight back to desirable levels. This is particularly applicable to lambs or weaners that were not fed supplements when grazing with their mothers.

Adult sheep, which are above these starting weights, can be allowed to lose some weight and condition at the start of a drought. This weight loss should be controlled. A drop in weight of 5 kg over a number of weeks and a drop back to store condition will save a lot of feed.

Plate 2. Feeding crushed maize grain

Using the rule of thumb based on condition scores, stop feeding gradually when less than a quarter of the stock remains at a fat score of 2 or less, after the drought breaks. In previous droughts, many breeders have experienced their heaviest losses during the period immediately following drought-breaking rain. Prolonged wet conditions turn sheep off their feed. In most circumstances, sheep are kept confined to restricted feeding areas until adequate pasture is available. At that point, allow increasing grazing time each day until full grazing is provided after 6 to 7 days. Allowing immediate full grazing will lead to digestive disorders.

Other Management Decisions

Shearing, pregnancy and lactation all increase the nutritional requirements of the sheep flock. Changes to mating, weaning and shearing times can sometimes be used to reduce feed demands during the drought.

The cost of drought feeding a breeding ewe for six months (including late pregnancy and lactation) is about 50% more than for a dry ewe, so savings can be made by delaying joining or by not joining. However, the long term effect of this action needs to be carefully considered.

Delaying the time of joining for an early autumn lambing flock has the potential to greatly reduce supplementary feeding costs. Joining may be put back a few weeks or changed to a late winter/spring lambing.

Although a change in the time of joining or reducing the number of ewes joined may save money or supplementary feed in the short term, this benefit must be weighted up against long term costs. Factors such as complicating management, disrupting the flock structure and altering future income levels need to be considered.

Early weaning, if not already practised, can reduce feed costs and simplify management of both the ewes and the lambs. The ewes can then be managed as dry stock and the lambs given priority (Jakhmola, 2008).

Crop Residues

Crop residues are a major source of roughage to Indian cattle. Fibrous crop residues like rice and other cereal straws are the major feed resources of the country. Even though the livestock feedstuffs are in short supply, still most of these crop residues are not effectively utilized. Usually in many parts of the country these materials, especially rice straw and other cereal straws, are wasted as they are fed in long form without chaffing. Further, the methods developed for improved utilization of these low-quality crop residues are not commonly practiced by the farmers as the advantage of these methods have not been demonstrated at the farm level. Hence, the nutrients from these crop residues are also wasted to a great extent. In order to utilize effectively the available crop residues, the methods for improved utilization of low-quality roughages like chaffing, ammonization with urea, ensiling, supplementing with leguminous fodder, enrichment with deficient nutrients, etc., should be demonstrated in each village so that the farmers can be convinced of the advantages (Mathur, 2004; Ramana *et al.,* 1990).

Agro-industrial By-products

The by-products obtained from grain processing (brans), oil seed processing (oil meals), pulses processing (chunni) are the major and important feed ingredients that are commonly fed to Indian livestock as concentrate feeds. Almost all the quantities of these by-products produced in the country are utilized as livestock feed. These by-products are considered as traditional or conventional feedstuffs. The brans are the main source of concentrate feeds accounting to more than 40% of the total concentrate feeds produced in the country. Another important source of concentrate feeds are oil meals constituting more than 20% of the total concentrate feeds produced. Thus, these two industrial by-products cover more than 60% of the total concentrate feeds fed to Indian livestock.

Conservation of feed stuff

Considering the difficult situation in feedstuffs availability in the country there is an urgent need for the conservation of feed stuffs for obtaining optimum output by feeding balanced rations to the superior stocks, cross-bred and indigenous breeds and their followers. The low-producing, uneconomic stock and the unproductive animals can be maintained on subsistence allowance till they are eliminated. Popularization of feeding balanced diets will reduce wastage of nutrients and ensure economy. Reduction and elimination of worm burden and shortening of dry periods is an effective way of feed utilization.

REFERENCES

Livestock Census (2003). Ministry of Agriculture, Government of India, New Delhi.

Jakhmola, R.C. (2008). Interventions to sustain small ruminant productivity during drought conditions. In: Small Ruminant Production in India - Strategies for enhancing (Eds Karim, S.A., Tripathi, M.K., Chaturvedi, O.H. and Shinde, A.K.) Satish Serial Publishing House, New Delhi, pp 193-209.

Mathur, B.K. (2004). Drought proofing through utilization of low – grade roughage and non-conventional feed resources for small ruminants. National Seminar on Opportunities and Challenges in Nutrition and Feeding Management of Sheep, Goat and Rabbit for sustainable production. (February, 10-12). Organized by *Indian Society of Sheep and Goat Production and Central Sheep and Wool Research Institute* avikanagar-304501, Rajasthan. *pp. 191-201.*

Narain, P., Sharma, K.D., Rao, A.S., Singh, D.V., Mathur, B.K. and Ahuja, Usha Rani (2000). *Strategy to combat Drought and Famine in the Indian Arid Zone.* Central Arid Zone Research Institute, Jodhpur, 65 p.

Ramana, J.V., Krishna, N., Parthasarthy, M. and Prasad, J.R. (1990). Effect of urea treatment on the protein degradability of pearl millet straw *Indian Journal of Animal Sciences* 16 : 1022-1023.

20

Extension Interventions in Transfer of Animal Husbandry Technologies relevant to drought prone areas

M.K. Mandape

Zonal Coordinator Unit, ZCU-VI (ICAR-Transfer of Technology), CAZRI Campus, Jodhpur 342 003

India is characterized by drylands/rainfed areas that constitute about 70% of the available land. These vast rainfed drought prone areas contribute 42% of food grains and about 75% of pulses and oilseeds. Out of total 588 rural districts, about 128 districts of the country fall in low to medium rainfall zone. Majority of these rainfed districts face great instability in crop production due to erratic and inadequate rainfall, further recurrence of drought every after 2 to 3 years is a common phenomenon. In this situation of uncertainities in agricultural production in rainfed farming, livestock production is appreciated to provide sustainability and economic stability to the farmers. Consequently, livestock production has been adopted by the farmers of these areas to augment and stabilize their income. Further, in these areas, crop residues are abundantly available to supplement the feeding of their livestock. Hence, role of livestock production in drought prone is manifold and due importance needs to be given for its development programmes in the areas.

Location specific studies in these areas have revealed the effectiveness of livestock farming in supplementing the income of the farmers. Further, the production and productivity of bovines, sheep and goat, poultry etc. in different agro-climatic conditions is most encouraging for generating additional income of the poor farming community. Livestock is a rapidly growing sector of Indian economy. It contributes 5.22% of Indian GDP and 28% value of output of agriculture and allied sector. The country is well endowed with livestock population. India is the highest producer of milk with 91 million tonnes. Milk production has grown at the rate of 3.4% in last few years and productivity has grown at the rate of 2.6% in last 10 years. The productivity of milch animal is 987 kg per year as compared to world average of 2200 kg/year and per capita availability of milk is 231gm/day as compared to world average of 270 gm/day. The value of milk output and its products is 70,000 crore rupees and that of dairy industry as whole is 1,05,000 crore rupees. The small ruminants in these areas are very important species of livestock playing important role in national economy. In rainfed areas, this livestock enterprise is advantageous due to their short generation intervals, higher rates of proficiency and the ease with which they can be marketed for meat, milk, skins and wool. This enterprise is associated with social and cultural fabric of millions of resource poor people of dry land areas. The small ruminants are very useful in semi-arid and arid zones where they can sustain themselves on sparce vegetation and extreme climatic conditions. Early sexual maturity, low age at first kidding (10-14 months) and multiple births in goat and sheep are the factors contributing to a rapid rise, specially in goatery.

Some facts of rainfed agriculture and livestock production.

The two states of ICAR Zone VI comprising of Rajasthan and Gujarat fall under low rainfall and low irrigation group based on average annual rainfall and % irrigated cropped area —

Particular		Rajasthan	Gujarat
# Average annual rainfall(mm)		494.0	877
# Percent irrigated cropped area		21.30	28.89
# Livestock population (in '000)			
(For 1982)	Cattle	13505	6994
	Buffalo	6043	4443
	Goat	15479	3300
	Sheep	13431	2357
	Total livestock	49651	17468
# Livestock Density			
(Heads/100 ha cropped area) (1982)			

Cattle	73	64
Buffalo	33	41
Goat	84	80
Sheep	73	22
Total livestock	270	160
# Growth rate in Breedable cows (1956-82)	0.25	0.23
# Growth rate in Breedable buffaloes "	2.10	1.93
# Growth rate in young stock - Cattle "	0.27	0.15
Growth rate in young stock - Buffalo "	1.90	1.58
# Yield per animal in milk (kg/day) Cow (1986-87)	2.720	2.599
" Buffalo	3.850	3.561
# Milk production ('000 tones)(1987-88)	4000	2997
# Per capita per day milk availability (gm) (1987-88)	270	213
#Growth rate in milk Production(%)	3.70	5.11
#Percent area under fodder crops to total cropped area	14.15	9.32
#Percent irrigated area under fodder crops	9.82	12.53
#Permanent pastures and other grazing land to Total cropped area	10.15	8.75
#Annual feeds and fodder availability during 1988-89 Green Fodder	70450*	25375*
*refers to 1985-86 Sugarcane Tops	69	791
Dry fodder	16512	8974
Oil seed cakes	1372	2613
Total Bran	246	130
#Fodder and feeds availability for bovines (Per capita/day) – 1988-89 (grams)		
-Green Fodder	9134	5633
-Sugarcane Tops	8	175
-Dry Fodder	2143	1992
-Oilseed cakes	203	132
Total bran	30	28
#Density of livestock population		
-Per ha of cropped area	7.31	5.77
-Per ha of geographical area	3.70	3.28
-Area/head	1.44	0.51

Arid (Low rainfall) and Semi-arid (Medium rainfall) areas are characterized for mono-cropping, mixed farming system with low value crops, grazing animals as well as cut and carry grass/fodder feeding livestock. Efforts are being made to transfer relevant technologies to the farmers of these areas, however, response is not satisfactory because of the constraints in livestock rearing such as low productivity, poor health and infectious diseases, less availability of feed and fodder, inconsistent breeding policies, a huge unproductive stock and restricted market access. The factors contributing to low productivity in animals are very slow rate of increase in population growth of crossbred animals, i.e. 19% of total population, non-availability of quality semen and/or breeding bull, poor conception rate following AI, lack of policy initiatives, scarcity of land for fodder cultivation, high feed deficits and lack of capital investments.

Technical and Extension Interventions Needed

Undoubtedly, in drought prone areas, livestock sustains poor farmers in times of natural calamities. Therefore, suitable technologies for sustaining animal husbandry need to be evolved. Such technologies could be related to clean milk production, rearing of healthy calves, disease prevention, optimum nutrition, enhancing the quality of dry fodder, storage of products, maintaining productivity during calamities, better reproductive efficiencies in animal etc. We need to have package of practices specific to locale and clients. Some of these technologies are available but have not reached farmers. Thus evolving region specific technologies assumes significant importance. Documentation, screening and preparing compendium of technologies is required. Further, many technological innovations require to be delivered at the farmer's door step, eg. AI service, vaccination, therapeutics – etc, accordingly extension approaches are to be followed. There is also need to recognize innovations by farmers and also to document their indigenous technical knowledge related to animal husbandry and wherever necessary validate the same and blend with modern technologies for wider acceptance. There is also a need of having interface between progressive farmers and other farmers of these areas for efficient propagation of technologies. Therefore, there is a need to restructure the technology dissemination from individual to collective, particularly in drought prone areas because common property resources and open access resources play a major role in animal husbandry practices.

The importance of animal husbandry is not often adequately reflected in Government programs. The lessons learnt from earlier programmes reveal that true potential of Indian farmers is realized through programme like operation flood which is a classic example of fusion between technological intervention and linkage with end users. Further, innovative technologies have the potential to revamp the growth of animal husbandry. It is also to be noted that technologies not suitable for small holder system are unlikely to gain wider acceptance. For appropriate extension interventions, resources, skills and opportunities of target groups need to be understood and analyzed in true perspective. A fresh and exclusive animal husbandry extension infrastructure is required involving state animal husbandry department, existing KVKs, NGOs, Private Industries and Farmers.

The lessons learnt through implementing several field extension education programmes at Indian Veterinary Research Institute, Izatnagar suggest that there should be comprehensive programme in animal husbandry targeted towards livestock owners. The list of activities could be manifold, like organizing animal health-cum-production camps wherein cases of clinical, gyanecological and surgical nature are treated by the scientists with close collaboration of field veterinarians; organizing deworming and mass vaccination campaigns with support of state animal husbandry departments, NGOs, Private Industries (Pharmaceuticals) and farmers; conducting joint diagnostic surveys for better animal health, organizing interface between scientists, field veterinarians and farmers, conducting demonstrations in livestock enterprise and on-farm-testing for providing better technological options and also for refinement of technologies involving scientists, KVKs subject matter specialists and livestock owners; imparting training to livestock owners, farm women and rural youth on relevant package of practices in animal husbandry with emphasis to undertake an economic activity, provision of farm literature and use of mass media for educating livestock owners about improved animal husbandry technologies. Concerted efforts by all concerned in the development of animal husbandry certainly will pave a way in enhancing production and productivity in animal husbandry and in harnessing the benefits at large by the farming community, more particularly by the resources poor of less endowed areas.

REFERENCES

Jana, B.L. (2005) "Appropriate technologies for drought prone areas", ATPA, Udaipur. Proceedings of First National Conference on KVKs held on 27-28 October, 2005 at New Delhi, Published by Division of Agril. Extension, ICAR, New Delhi.

21

Potable and hypersaline water use efficiencies in arid livestock

M.S. Khan

Central Arid Zone Research Institute, Jodhpur - 342 003

In order to survive and reproduce in an environmentally hostile arid region frequented by moderate to acute droughts in almost alternate years, an animal must possess the necessary physiological skills to deal with varying deficiencies of potable water and feed resources; wide spread hyper saline water coupled with high thermal heat load or develop skills as acclimatization progresses. A breed originated in a rather humid region with less water use efficient system and coping system to water and hyper saline water should not be expected to suddenly do well in such arid region. Before searching an efficient and appropriate physiological make-up of animals for arid region, it is worth mentioning here that poly functionalism exists in respect of body water, energy and protein turnover rate efficiencies in sheep and goats, which will allow considerable selection of the most hardy low producing animals for the drought prone arid region. It is now well established that animals, which have high rate of water use also have high rate of energy and protein use and *vice-versa*. Hence, animals may be reared for two purposes - (I) Low producing animals (drought hardy) with low water, energy and protein requirements, and (II) high producing animals (drought susceptible) with high demand of water, energy and proteins.

Environment and Animal

In the arid region of western Rajasthan average air temperature during the peak summer and winter may vary between 45-

48°C and -5 to 18°C with 12-18% minimum relative humidity of the air. The region also has high wind velocity during summer and monsoon, ranging from 13-14 km hr^{-1}. Scanty and erratic precipitation and a high rate of evapo-transpiration lead to insufficient recharge of the underground water and supplies. Soil and water salinity is also moderate to high at several places in the region leading to high rate of mortality of animals due to nitrate poisoning and incidences of flourosis.

In any case all livestock, except high yielding dairy cattle and buffaloe, which enjoy the care and attention of their human masters, need to graze or browse for an average of 6-8 hours per day on community range under the sun during peak summer. Since it is not possible to convert this harsh environment to a hospitable one where an animal may express its full genetic production and reproduction potential, it is during this period that their claim to desert habitation is tested.

Heat and water relationship

Most of the livestock are not able to store heat during day time and dissipate it during relatively cool evenings and nights by physical means i.e., radiation, conduction or convection. Sheep and cattle in general are thermo stable and depend mainly on the expensive water evaporative cooling mechanism. Camel and to some extent black Marwari goats, are thermo labile and dissipate most of their heat load during cool hours of the evening and nights mainly by radiation. A 500 kg camel and 30 kg Marwari goat can store about 2500 kcal and 830 kcal in a day when their body temperature increase by 6°C and 2°C, respectively. The camel and marwari goat do this very conveniently to save the precious water.

As regards to drinking water, large variations are observed among individual animals in their water requirements. Such differences are influenced by nature of the feed consumed, environmental temperature, humidity and physiological conditions of the animals. Adding proteins, minerals and salts to the diet may increase water requirements as the body must expel the metabolites and excess salts through the kidneys by way of urine.

Under arid conditions, livestock in general drink water about double the quantity of the air dried feed ingested by them. On dry matter basis, sheep, goat and cattle usually consume feed at the rate of 2.5% of their body weight day^{-1}. Thus, the daily water intake of these animals comes to about 5% of their live weight. The normal average water intake of Marwari goats during summer, winter and the post-monsoon season has been recorded as 8.3, 6.3 and 5.6 l 100 kg^{-1}, respectively (Ghosh and Khan, 1980). African goats consume water at the rate of 8% of their body weight day^{-1}. Camel and donkeys drink water at the rate of 5 to 6% and 8 to 10% of their body weight during winter and summer, respectively. In some areas where dew is more common and heavier, it could be an important source of free water for livestock grazing on such pastures.

Rise in environmental temperature from 16°C to 30°C at 60% to 70% relative humidity led to increase in daily water intake, evaporative water loss, urinary out put and total body water in Friesian bull (Shebatia et al., 1992). The water intake has been found to be significantly increased to 16-20% after 6 km grazing in goat and sheep during summer (Khan and Ghosh, 1989). Water intake also increased significantly in lactating sheep (El-Sherif and Hassanaein, 1996) and on high salt load (El-Nouty et al., 1988, Abdel-Samee, 1996).

Maintenance of body weight during absolute and intermittent water deprivation

Acute water restriction

Of the various physiological responses during water stress as revealed by various experiments is the loss in body weight of dehydrated animals. During summer (34-38°C) the Marwari goat, on an average, lost only 1.5% of its body weight per day while the Marwari sheep lost as much as 6% per day. In another study when the Marwari sheep was put on 2 days dehydration regime during summer, it had lost its body by 11% on its first day of dehydration and by 6% on the second day of dehydration. Bedouin goats had lost 32% (on an average of 8% day) of their body weight after 4 days of dehydration. In another study, however, when absolute water deprivation was imposed for 4 days at an environmental temperature 35°C on goats weighing about 40 kg, the animals lost about 3% of their body weights per day (Khan *et al.*, 1979). The secret to dehydration tolerance in lean animals, however, seems to be because of their initial higher body water contents, which is perhaps used for various body functions during dehydration. It seems that young animals with low body weights and fat contents, but higher body water contents put more resistance to water deficiency conditions than the old animals with higher body weights, fat contents and lower body water contents. Reports from Australia indicate that during water deprivation, shorthorn cattle lose about 8% of body weight per day; Merino sheep about 4-5% per day and camels about 2% per day. This would put the Marwari goats in a category very similar to that of the camel in respect of body weight maintenance during absolute water deprivation.

Partial water restriction

When maintained under partial water restriction (1/4 of the normal) schedule during one year period (summer to summer), the Marwari goats usually gain in body weight. It has been estimated that with a flock of 100 adult goats about 171 l of water/month may be saved, or 400 goats may be kept on the water ration of 100 without any ill effect, by limiting the daily intake of the animals to a quarter of what they would normally drink.

Intermittent water restriction (Watering after every 4th day)

After the monsoon till the next summer, the Marwari goats receiving water after every 4th day did not show any ill effect either on body weight or on their general performance. The percentage increase in body weight of Marwari goats over the period was 59% and 48% for the daily *ad lib.* watered and after 4 days *ad lib.* watered goats, respectively. The average water consumption of animals watered after 4 days was about 3 l/100 kg b.wt./24 h, while the daily watered goats had an average intake of about 6 l/100 kg/ b.wt./24 h. Apparently these animals, when watered after every 4th day, perform better than the Marwari sheep, which under water restriction reportedly lost, on an average, 6% of their body weights per day. However, it has been observed that intermittent (twice a week) watering in Marwari and Magra sheep does not in anyway affect their body growth or productivity (Ghosh and Abichandani, 1981), besides considerable saving of drinking water.

Watering after every 5th day

Effect of intermittent watering (watering *ad lib.* after every 5 days) on the maintenance of body weights of Marwari goats and Marwari sheep was studied for 3 years. During winter season, the Marwari goats watered *ad lib.* and after every 5 days registered no losses in their body weights. However, the same animals lost their body weights to the extent of 10-12% during summer but their feed consumption remained unaffected. On the other hand, the Marwari sheep watered *ad lib.* after every 5 days, not only lost their body weights during summer (about 20%) and winter (about 10%) but their feed consumption was also decreased during these two seasons, though less during winter, but more during summer. When compared between sheep and goats, the sheep consumed more drinking water (about 29% during summer and 22% during winter) than the goats, when maintained on a schedule of watering after every 5 days. However, neither the sheep nor the goats could tolerate water restriction for 5 days when the environmental temperature rose beyond 40°C and therefore, this schedule of intermittent watering requires constant vigil on the environmental temperature while maintaining these animals on such intermittent watering schedule.

Body water distribution pattern

The distribution of water in various compartments in normally watered desert animals has been reported by several workers (Kataria *et al.*, 2003, Hammord *et al.*, 1990, Shkolnik and Choshnaik, 1985, Khan and Ghosh, 1989 and Dunshea *et al.*, 1990). The Marwari sheep and Marwari goat have almost similar plasma volume (PV) and total body water (TBW). However, extracellular fluid volume (SCN space) - a very important water compartment for combating water stress is significantly higher in the Marwari goats than in the Marwari sheep. In the Beduin goat PV and TBW are significantly higher than in Marwari sheep and Marwari goat, but SCN space is lower in the Beduin goat than in the Marwari goat. The Indian desert camel, however, has higher levels of PV, Cell and Gut water and TBW than in the Marwari sheep and Marwari goat. Such species or breed differences with regard to various body fluid compartments may be attributed to the inter-species differences in the mechanism of body water regulation during water deficiency conditions.

Physiological manoeuvring under insufficient water supplies

For any desert animal, the maintenance of circulatory volume is of great importance under water deficiency conditions because of two reasons: (i) to avoid circulatory failure, which may occur if the blood gets thick and, (ii) to dissipate the internal body heat through the circulatory medium. During water restriction, water lost from the body of an animal is drawn from various body water compartments and the degree to which these compartments are depleted during dehydration differs from species to species. For example, in Marwari goat 4-days of water deprivation during summer lowered the plasma volume (PV) by 13% of normal. In Marwari sheep, on the other hand, PV was reduced by 43% under similar environmental conditions (Table 1). The Australian Merino sheep also lost about 45% of PV after 5 days of water deprivation. The actual mechanism involved in the retention of plasma water in the goat (Khan and Ghosh, 1985) and also in the camel (Kataria et.al., 2003) appears to be associated with the retention of plasma proteins, particularly, albumin, in the vascular bed.

Table 1 Body fluid components (Mean± SE) in normally watered and water restricted Marwari goats

Parameters	Body water (l) Before water Restriction (Mean± SE)	After day 4- of water Restriction (Mean± SE)	Change of complete water restriction from *Ad-lib.* value (%)	Water (l) lost (Mean± SE)	Percent of the total body water lost
Body weight (kg)	40.10 ± 0.75	35.30± 1.08	(-) 11.97	-	-
Total body water(1)	24.46 ± 1.53	18.21± 1.09	(-) 25.55	6.24 ± 2.11	---
Plasma volume(1)	1.47 ± 0.03	1.28 ± 0.05	(-) 12.92	0.19 ± 0.07	3.04
Blood volume (1)	2.03± 0.07	1.93 ± 0.06	(-) 4.92	0.14 ± 0.05	2.24
Extra-cellular fluid volume (1)	11.49± 0.22	10.62± 0.32	(-) 7.57	0.90± 0.28	14.42
Cell and gut water (1)	12.96 ± 1.52	7.59 ± 1.03	(-) 41.43	5.58 ± 1.80	9.42
Interstitial fluid volume (l)	10.02 ± 0.19	9.33± 0.31	(-) 6.88	0.71± 0.27	11.37

Unlike Macfarlane's Merino and the Marwari sheep, which had severe reduction in the SCN volume at the end of dehydration regimes, the Marwari goat tends to conserve its SCN volume some what more efficiently, there being a reduction of SCN volume by only about 8% (Khan, *et. al.,* 1979). In the Marwari goat the reduction in the cell and gut water is of the order of 41% at the end of 4-day dehydration regime, while in sheep it may be to the extent of 30% only. Cell and gut water is of special significance during periods of water stress in the Marwari goat because it is this water, which is mainly relied upon by the animal for maintaining its normal circulatory volume (Khan, *et al,* 1979). The cell and gut water may, thus, be considered as "Emergency Water Reservoir" which is used to avert the circulatory failure in the goat. The inability of sheep to use this water fully in times of water crisis and its primary dependence on circulatory water for evaporative cooling and other essential purposes, makes the sheep decisively less desert worthy than the goat.

When deprived of drinking water for as long as 13 days during late winter, Marwari goat showed an extraordinary ability to maintain its body fluids, whereas Marwari sheep almost reached its limit of physiological tolerance by the end of this period. The rate of body water loss in the goat, sheep and the donkey was 46, 87 and 99 ml $kg^{-1}day^{-1}$, respectively, during 13 days water dehydration regime. During dehydration regime, a rank order in respect of maintenance of SCN volume in these species has been noted as : Marwari goat < Marwari sheep < Desert donkey. Of the total amount of body water lost, the cell and gut water of the sheep contributed to the SCN space was 24% only in contrast to 80% in the goat (Khan *et al.*, 1979) and the camel (Macfarlane *et al.*, 1963 and Kataria *et al.*, 2003). Interestingly, it is the interstitial fluid volume (IFV), which contributes about 60% to the depleting extra cellular fluid compartment in sheep.

Body water and its turn over rate

Total body water in Marwari sheep, Marwari goats and Parbatsar goats has been found to be 52.9%, 58.5% and 61.6% of body weight, respectively. In Sahelian goats, Maradi goats and Bedouin goats (Shkolnik and Choshnaik, 1979) it was 77.8%, 66.1% and 76.3% of body weight respectively. TBW in Uda and Yankasa sheep, Merino, Awassi and Ogaden sheep (Macfarlane, 1982) was 80.5%, 73.5%, 63.9%, 71.5%, 73.5% and 68.0% of body weight respectively. It appears that Indian desert sheep and goats have significantly less TBW in comparison to exotic sheep and goats inhabiting the arid region.

The data revealed considerable variation in body water turnover rates (BWTRs) in different breeds of Indian desert sheep and goats. The Marwari sheep turned over nearly 17 per cent more water per day than the Marwari goats. Parbatsar goats and

Marwari sheep, however, have more or less similar body water turnover rates i.e., 185.1 and 172.0 ml/ $kg^{0.82}$/day. Published reports generally ascribed lower BWTRs in desert adapted goats like, Sahelian goats and Maradi goats (Aganga et al., 1989), Beduin goats and Baladi goats (Shkolnik, et al., 1979) than the desert sheep like, Barki Rahmani sheep (El-Nouty et al., 1988) and Marwari sheep (Khan and Ghosh, 1983). The body water turnover rate in Marwari goat (147.0) was significantly lower than the Parbatsar goat (185.0), Somali goats (185.0) and East African goats (319.0).

Young Marwari and Magra lambs, Marwari and Parbatsar kids have high total body water than their adult counterparts. Young Ungulates and Merino lambs have also been found to have higher total body water than their adult counterparts. In other words, total body water per unit body weight decreases with age. The progressively lower water content during the growth period is probably related to increased synthesis of connective tissues that are less hydrated as compared to muscle tissues. Lean tissue contains 72% water, whereas adipose tissues have only 20% water. BWTRs were also significantly higher in Marwari lambs and kids than their adult counterpart. A similar age effect in respect of BWTR has been reported for Merino sheep. Female Marwari kids tend to have higher rates of water turnover than male kids do.

Table 2. Body water and its turn over rate in normally hydrated, dehydrated and rehydrated animals

Species	Parameters	Normal (Mean ± S.E.)	Dehydration (Mean ±S.E.)	Re-hydration (Mean ± S.E.)
Marwari goats (4)	Body water (% B. Wt.)	±1.75	±1.26	±1.72
	BWTR (ml $kg^{0.82}$ day^{-1})	116.41± 2.03	11.59± 0.65	124.22± 2.66
Marwari sheep (4)	Body water (% B. Wt.)	±0.93	±0.06	59.29±0.88
	BWTR (ml $kg^{0.82}$ day^{-1})	120.21± 2.14	12.47± 0.70	146.72 ± 5.68
Marwari donkey (2)	Body water (% B. Wt.)	±4.35	±0.62	±1.48
	BWTR (ml $kg^{0.82}$ day^{-1})	170.34 ± 11.68	21.32± 0.37	170.34 ± 11.68

Among Indian desert livestock species, a significant ($p<0.05$) rank order was observed in respect of water turn over rates in hydrated and rehydrated sheep, goat and donkey (Table 2). It was lowest in Marwari goats and highest in donkeys; Marwari sheep occupied a middle position. Positive correlation between BWTR and heat load has also been observed in sheep, goat and donkey of the Indian desert (Khan and Ghosh, 1983) and Oryx and Eland antelopes of the African arid zone. Body water turn over (BWTR) estimates suggests that the Marwari sheep turned over nearly 17% more water per day than the Marwari goat. The Parbatsar goat also of the Rajasthan desert, however, behaved similar to the Marwari sheep in this respect. The rate of body water turn over in the Marwari goat would appear to be lower than that reported for goat breeds from other dry regions like, Australia and Israel.

Hypersaline water use

Problem of high nitrate and fluoride in underground drinking water is acute in western Rajasthan. 60-64% wells have exceeded permissible limits of total soluble salts, fluoride, nitrate, chloride and sulphate contents. Consequently intake of such hyper saline water spread in large areas of arid region often leads to high mortality of animals, particularly during summer. Therefore, it is imperative to ascertain adverse effects, if any, of high intake of fluoride, nitrate, chloride and sulphate, the main constituents of saline water, on the overall productivity of animals and also on kidding/lambing and their growth rates.

The experiment on the effects of hyper saline water intake was started with synthetic saline water containing nitrate (8 g), fluoride (24 mg), chloride (6 g) and sulphate (2.4 g) animal^{-1} day^{-1} in sheep and goats for 1 year period. After having recorded its safe limit the concentration increased to 30 mg fluoride + 6 g nitrate + 7 g chloride + 15 g sulphate animal^{-1} day^{-1}. Later on the concentration of salts increased to 60 mg F, 12 g NO_3, 7 g Cl, and 18 g SO_4 animal^{-1} day^{-1} which has been found to be safe to desert sheep and goats. No adverse affects either on body weight of adult animals, body growth of lambs and kids or blood constituents like plasma proteins, albumin, globulin, A/G ratio, % PCV, Hb, inorganic phosphorus, alkaline phosphatase and Methomoglobin were observed. Here after the concentrations of these salts were increased to 80 mg F + 16 g NO_3 + 10 g Cl + 20 g SO_4 animal^{-1} day^{-1}. At this level of salt concentration also no adverse effects on body weights of adult sheep and goats, birth weight of kids and lambs; and body growth rates of these lambs and kids have been observed. Also no stillbirth, abortions were recorded in saline treated animals.

It is concluded that intake of 80 mg F + 16 g No_3 + 10 g Cl + 20 g So_4 animal^{-1} day^{-1} is safe to Marwari sheep and goats.

REFERENCES

Abdel-Samee, A.M. 1996. Egypt. J. Agric. Sci. (UK), 97(2):137-147.

Aganga, A.A.; Umunna, N.N.; Oyedipe, E.O. and Okoh, P.N. 1989. *J. Agric. Sci.* (U.K.), 113 (2): 255.

Dunshea, F.R., Bell, A.W. and Trigg, T.E. 1990. British J. Nutr. 64(1):121-131.

El-Nouty, F.D.; Hassan, G.A.; Taher, T.H.; Samak, M.A.; Abo-Elezz, Z. and Salem, M.H. 1988: *J. Agric. Sci.* (U.K.), 111(1), 27.

El-Sherif, M.A. and El-Hassanein, E.E. 1996. Alexandria J. Agric. Res. 41(3):1-9

Ghosh, P.K. and Khan, M.S. 1980. *The goat in the desert environment*, CAZRI Research Bulletin, Jodhpur, No 12, pp 1-25.

Ghosh, P.K. and Abichnadani, R.K. 1981. *Water in the eco-physiology of desert sheep*. CAZRI Monograph No.17. pp 1-42.

Hammond, A.C., Waldo, D.R. and Ramsey, T.S. 1990. J. Dairy Sci. 73(11): 3141-3145.

Kataria, N., Kataria, A.K., Agarwal, V.K., Garg, S.L. and Sahani, M.S. 2003. J.Camel Practices and Research, 10(2): 99-105

Khan , M.S. and Ghosh, P.K.1983. *Journal of Arid Environment* 6:173-175.

Khan, M.S. and Ghosh, P.K. 1989. *Ind. J. Anim. Sci.* 59(5):600-603

Khan, M.S. and Ghosh, P.K. 1985. *Ind. J. Animl. Sci.*, 55(3) :169.

Khan, M.S., Sasidharan, T.O. and Ghosh, P.K. 1979. *Experientia* 35: 1185-86.

Macfarlane, W.V. 1982. In: Proc. 2nd Inter. Conf. Goat Production and Disease, Arizona, p. 375.

Macfarlane, W.V.; Morris, R.J.H. and Howard, B. 1963. *Nature* (London), 197: 270.

Shebaita, M.K., Yousri, R.M. and Pfau, A. 1992. *Intern. J. Anim. Sci.*, 7(2): 235-240.

Shkolnik, A. and Choshniak, I. 1985. *Physiological responses and productivity of goats. Stress physiology in livestock*. Vol. 2 Ungulates, pp 39-57, (Ed.) Yousef, M.K. CRC Press, Inc., USA.

Shkolnik, A., Maltz, E. and Choshniak, I 1979. 5th Inter. Symp. on Ruminant Physiology, Clermont, Ferrand, September, 1979.

22

Role of Probiotics in Improving Livestock Productivity

P.P. Rohilla

Regional Research Station, Central Arid Zone Research Institute, Pali-Marwar (Rajasthan) - 306 401.

Introduction

There is growing interest in probiotics, applied as therapeutic and preventive agents, such as *Lactobacillus*, *Bifidobacterium*, *Pediccoccus*, *Leuconostoc* and *Enterococcus*; some strains of yeast (*Saccharomyces Candid*) and mould (*Apsergillus oryzae* and *niger*) are used most often in the production of probiotics.

What are Probiotics

Probiotics are defined as viable cell preparations or foods containing viable bacterial cultures or components of bacterial cells that have beneficial effects on the health of the host. Hence, live microorganisms and substrates (sometimes known as prebiotics) that promote growth of indigenous probiotic microorganisms and components of probiotic microorganisms are included.

Probiotics preparations have a beneficial influence on animals, manifested by higher body weight gains, increased activity of saccharase and maltase. These are health enhancing functional food ingredients used therapeutically to prevent diarrhoea, improve lactose tolerance, modulate immunity and antineuoplastic effects. They also have potential to prevent cancer and lower serum cholesterol levels. e.g. *Lactobacillus* and *bifidobacterium* etc. (Akman and Yagci, 2002).

Prebiotics

A prebiotic substance has been defined as a non-digestible food ingredient that beneficially affects the host by selectively stimulating the growth and/or activity of one or a limited number of bacteria in the colon. Therefore, compared to probiotics, which introduce exogenous bacteria into the colonic microflora, a prebiotic aims at stimulating the growth of one or a limited number of the potentially health promoting indigenous micro-organisms, thus modulating the composition of the natural eco-system.

Prebiotics (non-digestible oligosaccharides) can fortify intestinal microflora and stimulate their growth. Non-digestible carbohydrates act as selective prebiotic fermentation substances for beneficial colonic proboitic bacteria to produce short chain fatty acids. These short chain fatty acids elicit effects such as alteration in preneoplastic lesions, enzyme induction, suppression of mutations and binding of potential carcinogens. Prebiotic–Probiotic interaction and activity is therefore key role in colon cancer protection by non-digestible carbohydrates. Further, prebiotics have been found to stimulate absorption of several minerals to improve mineralization of bone by increasing the availability of Ca, Mg, Zn and Iron etc. Effect of prebiotics depend on dose, the time of administration, content of calcium in the diet, the part of skeleton investigated and the age of the concerned studied (Cashman and Tannock, 2002).

In other words Probiotics are commonly defined as viable micro-organisms (bacteria or yeast) that exhibit a beneficial effect on the health of the host when ingested. They are used in feeds, especially in fermented dairy products; whereas, Prebiotics are non-digestible foods which improve the growth of bacteria in colon. Intake of prebiotics can significantly modulate the colonic microbial flora by increasing the number of specific bacteria. Non-digestible oligosaccharides in general and fructo-oligosacchaides in particular are prebiotics (Apan, 2000). In short, a probiotic is a live/viable microbial feed supplement; whereas prebiotic is a non-viable food ingredient selectively metabolised by intestinal species seen as beneficial.

Importance of Probiotics

The health benefit of probiotics have been the subject of increased research interests, these food supplements have been demonstrated to alter the pre-existing intestinal flora so as to provide advantage to the host. The scientific evidence both for and against their role in promoting health and treating diseases is important, specific attention is turned to their effects on immuno-modulation, lipid metabolism, cancer prevention, diarrhoea, helico-bactor pylori, necrotizing, enterocolitis, allergy and inflammatory bowel diseases. A combination of probiotics and prebiotics may be the most effective strategy for maximising anti-carcinogenic effects (Titelbaun and Walker 2002).

Probiotic administration transiently alters the gut micro flora by donating cells to the ecosystem. It may be possible to use carefully formulated products in the alleviation of inflammatory bowel diseases or allergic diseases if the safety of such probiotics can be guaranteed.

In recent years increasing attention has been focussed on the possible beneficial effects of probiotics, such as enhanced resistance to invading pathogens, improved bowel function, anti-colon cancer properties, lipid lowering action, improved calcium bioavailability amongst others. The increased utilisation of bioactive components (pre and probiotics) in the formulation of food products that may have a significant effect on human health, especially on the prevention and treatment of different diseases. Probiotic bacteria are sold mainly in fermented food and dairy products play a predominant role as carriers of probiotics. These foods are well studied for promoting the positive health image of probiotics for several reasons.

Lactobacillus group of organisms are the main type in probio-tics and spores of these bacteria are commonly used for this purpose. When fed to animals and poultry, they enhance FCR leading to more production of better quality. Besides, there is a substantial fall of losses due to infection among them. These are given as feed additives or medicaments. These bacteria are mainly responsible for fermentation and they are seen in the fore-stomach of rumen of cattle, sheep and goats. They are needed for cellulose fermentation in it. In addition, they synthesize many vitamins of "B" complex group and also vitamin "K" in the GI tract. They also serve as a source of protein to their hosts after undergoing lysis. Thus, they are beneficial to the body.

These microorganisms can be classified broadly into two groups. The permanent flora that are normally seen in the gut.

These are useful ones and are the gut residents. The other group is temporary flora that are seen at times. They stay for sometime only and they exert their effects as they pass through the gut. Moreover, these friendly bacteria can also stimulate the growth and multiplication of the usual resident bacteria of the gut. e.g. a new *Bacillus subtilis* strain, which is not normally present in the gut, is found to be very effective in promoting growth and other desirable characteristics in animals and poultry when spores of this organism are fed to them. These changes can improve GI function leading to improved health and longevity.

In healthy animals and poultry there is a balance between useful and harmful bacteria in the gut. When this balance is disturbed and when harmful ones take upper hand discomfort or diseases result. Such animals tend to become weak. Their growth rate and production is low or reduced, besides deaths at times. On the other hand, when useful bacteria take upper hand health is promoted. The growth rate and production of these animals are bound to be at their maximum.

Probiotics/bacteria are said to produce some antibiotics like, Acidolin, Acidophylin, Lactin etc. These are highly effective against pathogenic bacteria viz., *Eschrichia coli*. the main harmful bacteria of the stomach and intestines. In addition, these bacteria can also produce certain organic acids like, Formic acid, Lactic acid, Acetic acid etc. and increase intestinal acidity. Thus, the contents of stomach and intestines are not conducive to the growth and multiplication of many pathogenic bacteria. This is not also favourable for coccidian, so very advantageous to chicks. Added to these effects, probiotic bacteria being aerobes lower oxygen tension and render gastro-intestinal canal less prone for the growth and multiplication of pathogenic bacteria that are mostly anaerobic. Hence, the counts of *Lactobacillus* group, the useful bacteria of the gut are increased. Consequently, the counts of *E.coli* and others that are harmful are decreased. Besides, they compete with pathogenic bacteria of the gut for the use of nutrients available in the stomach and intestines. Thus they deprive pathogenic bacteria of their nutrients to some extent.

In stress, the bacteria of the *Lactobacillus* group decreases to a great extent in the GI tract, resulting an increase in the counts of harmful ones leading to impairment of digestion or GI functions. These changes greatly affect health and activity in them. The growth as well as production in animals and poultry is reduced to a marked extent. When the stress is more and prolonged animals become weak and dull. They are more prone to diseases and deaths. There are many types of stress in animals and poultry. e.g. excessive heat, cold, medication with antibiotics or vaccination. Then, it is better to feed them with probiotics. This is natural way of combating diseases. It has been found that when animals are fed with spores of these organisms they recover more easily as well as quickly especially in ailments of GI tract.

Role in Livestock Production: Probiotics as Animal Growth Promoters

The existing nutritional and managemental practices are not sufficient for achieving higher animal productivity. In order to promote growth and to impart flavour to animal feeds additives are incorporated from external sources in the animal feeds. Minerals (Ca, Mg, P, Zn, Cu), vitamins (fat soluble as well as water soluble) and hormones constitute important groups of feed additives. Since a long time these additives had been used for control of infections. It is believed that use of antibiotics could enhance growth up to 20%. However, some of the antibiotics are not currently in use due to therapeutic purposes and their harmful side effects.

Unchecked and indiscriminate use of antibiotics has had their ill effects and developed resistant microbial strains, which made treatment of infectious diseases difficult. Residual effects of antibiotics led to development of allergy among consumers of animal foods like, milk, meat and other products resulting diarrhoea, enteritis and food poisoning etc. (Moro *et al.*, 2002).

In view of above, animal scientists have been engaged in finding suitable substitute for conventional antibiotics additives. For centuries curd and fermented milk were considered to help in controlling GI problems. Such foods are today regarded as valuable probiotic preparations. Their major advantage is that they are cheaper and don't exert any side effect like development of allergy and microbial resistant.

Probiotics unlike antibiotics facilitates growth of desirable microflora. This microflora in turn modifies the environment of GI tract and results in better utilization of feeds. Important microbes acting as probiotics include : *Lactobacillus* spp., *acidophilus, Lactobacillu bulgaricus* and *Lactobacillus salivarius* are widely used. Use of these strains is found to promote body weight, modify gut microflora and help in controlling cholesterol levels, some anti tumour activity also.

Probiotic research has gained substantial momentum in recent years. As a result several types of probiotics have become available as safe growth promoters to livestock owners. The probiotics are now available as gels, pastes, powders, capsules and granules etc. In some cases probiotics could even help remedy of clinical conditions involving GI tract. Probiotics have opened up a new area of study in improving livestock productivity. It is believed that their use would benefit farmers in great measure and help in reducing dependency on antibiotics. Their use will eliminate ill effects of antibiotics and will result in substantial economic gains to the livestock owners. That's why these are also known as "Future antibiotics".

Animal feed additives are used worldwide for many different reasons. Some help to cover the needs of essential nutrients and others to increase growth performance, feed intake and therefore optimises feed utilisation. The health status of animals with a high growth performance is a predominant argument in the choice of feed additives. The consumers more and more question the use of feed additives.

Therefore, the feed industry is highly interested in valuable alternative, which could be accepted by the consumers. Probiotics, prebiotics, enzymes and highly available minerals as well as herbs can be seen as alternatives. Herbs, spices and their extracts (botanicals) have a wide range of activities. They can stimulate feed intake and endogenous secretions or have anti-microbial, coccidio-static or anthelmintic activity (Wenk 2003).

The animals are constantly exposed to a diversity of health challenges and the gastrointestinal tract is a major site of exposure. The use of antibiotics in animal feed to improve growth is a common feature. The use of probiotics (biotherapeutic agents such as, *lactobascillus*, *bifidobacterium*, *bacillus*, *streptococcus*, *pedicoccus*, *entrococcus*, *saccharomyces*, *aspergillus* and *torulopsis*) in animals in relation to gastrointestinal tract and the immune system are of prime importance.

Probiotics or immune supplements can inhibit pathogenic gut microorganisms or make the animal more resistant to them. The administration of probiotics seems to be promising with regard to their capacity to modulate the bacterial composition in the colon and there are indications that probiotics can beneficially influence colonic metabolism. To recognise the potential advantage of probiotic use, we need to understand how bacterial cells function in the gut ecosystem and how bacterial functions impact on the host.

Research work has been done in different livestock species in our country and in abroad, some of them are being mentioned here as under:

Swine/Pig

The alternatives to antibiotics as growth promoters in pig nutrition like probiotics or a combination (of both probiotics and prebiotics i.e. synbiotics) are the research interests today. Pairing non-digestible-oligosacchairdes (NDO) and probiotics strains that have the metabolic potential of fermenting the supplied NDO at a competitive rate compared to the indigenous microflrora is likely to be a successful strategy in controlling the intestinal ecosystem. The expected benefits are in improved survival rate during the passage of the probiotic bacteria through the upper intestinal tract and a more efficient implantation in the colonic microbiota together with a stimulating effect of the NDO on the growth and/ or activity of both exogenous and endogenous bacteria.

Several workers in India and abroad (Christopher 1987; Karput and Pudenko 1996; Rekiel and Wiecek 1996; Spieler and Manner 1996; Tandani and Terreni 1996; Vassalo *et al.* 1997; Letellier *et al.* 2000; Naskar *et al.* 2000; Baynes *et al.* 2001; Zimmermann *et al.* 2001) have reported beneficial effects of probiotics in improving live weight, absolute gain in body weight, average daily gain and in reducing the mortality in grower pigs and promoted post-weaning gut health in piglets.

Probiotic microorganisms produce organic acids reducing pH of intestinal contents. In this way they inhibit the development of pathogenic bacteria, especially, the *Coli* group, which are the main reason for the diarrhoea in piglets. The probiotic controls the metabolic activity of the gut microflora, which leads better digestion and absorption of nutrients. Further, use of probiotics, vaccination and acidification of drinking water has been used successfully as means of reducing *Salmonella* in pigs.

Poultry/ Brolier Chicken

The digestive capacity of the chicken can be modelled according to interactions between feed-related and bird-related factors known to affect energy metabolism. But microflora have a highly significant impact on between bird variation in the digestive capacity of the broiler chicken. Poultry meat and eggs are important source of human pathogens. *Salmonella* is a major cause of human food-borne infections following consumption of poultry products.

Poultry are dependent up to their feed for their requirements of many vitamins belonging to "B" complex group and a few others to a major extent. When probiotics are fed to them, they can produce some of these in their bodies. Thus they promote health and longevity, as some of these essentials of health are reproduced in them. When fed as feed additives, they enhance GI function leading to improved feed efficiency and production. There is a substantial fall in losses due to infections, as their presence in the GI tract is not conducive to the multiplication of many pathogenic bacteria and coccidian. When given to sick birds or after a disease out-break in a form, they quickly regained their normal function and showed better appetite. This is more pronounced after intestinal ailments. They are useful to avert weakness at all times. Hence, feeding of probiotics is good to all types of poultry at all times.

When given as feed additives or as medicaments probiotics are safe and there is no evidence of toxicity or harm to the hosts in any way as this is nature's way of curing diseases. They pose no palatability problems, as they do not have any sort of taste or smell. They can be mixed up with feed or water easily and administration of medicaments or tonics etc.

Probiotic feeding in stress and in convalescence is helpful and useful. In ruminants, these are very much needed as they eat enormous quantities of grass and roughages. They promote digestion of these in their rumen. Some of these transient bacteria that are useful may get colonised in the GI tract after some mutation. Hence, there is a need to fed these probiotics to animals and poultry. There is an urgent need to know about probiotics fed animals in greater details.

In recent years a good number of probiotics containing different bacteria and mould cultures along-with other growth promoters are available in the market. Manufacturers are recommending different levels of supplementation in the poultry broiler ration. The manufacturers have frequently claimed that these improve growth rate, feed conversion and liveability thus resulting in economic broiler raising.

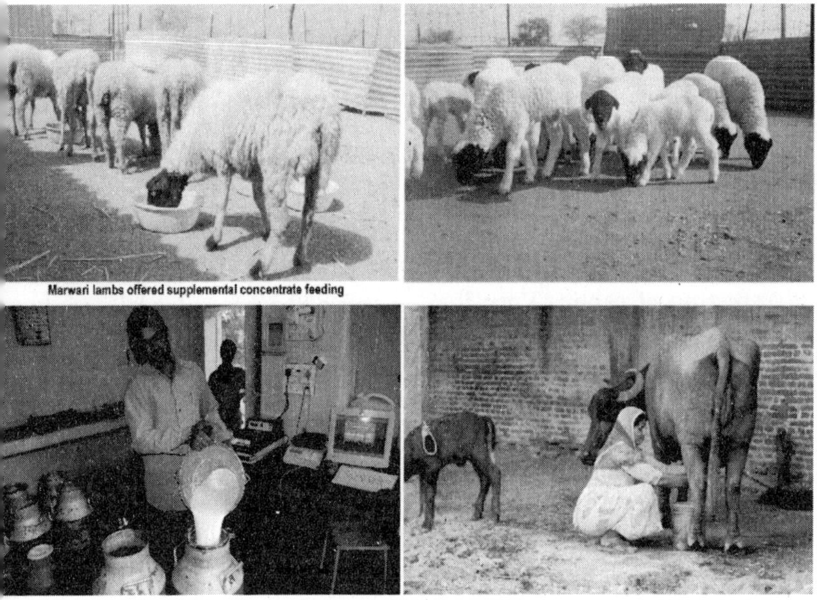

Marwari lambs offered supplemental concentrate feeding

Plate 1.

Different workers from India and abroad have reported on the use of probitotics in poultry broilers and chicken (Gandh and Nagarathanam 1990; Thapliyal 1993; Pradhan *et al.* 1998; Mandal *et al.* 2001; Giardini *et al.* 2001; Hughes and Corbett 2001; Maiorka *et al.* 2001; Immerseel *et al.* 2002) which helps in solving problems like gastrointestinal infections, low egg production, feed conversion ratio and contamination of eggs and carcasses.

The efficacy of different probiotics differ accordingly, might be due to variation in the composition of probiotics and/ or due to bacteria and growth promoters, however, the major variation has been due to bacteria. The superiority of some of the probiotics have been reported in terms of growth promotion and feed conversion resulting in better economy probably due to presence of *Lacto-bascillus* sporogenes bacteria. This bacterium due to spore forming has been found more resistant to the adverse conditions during the storage and in the gut also.

Role of probiotics in immune response of birds against bacterial, viral and protozoal infections has been established. Further, metabolites from intestinal fermentation can be absorbed and can significantly affect the organoleptic characteristics of poultry meat and fat. By using appropriate diets it is possible to change microbial metabolism and the ratio of the final metabolites produced. The ratios of intestinal microbial populations can also be affected by other factors such as stress and administration of substances like probiotics, prebiotics and antibiotics, which require careful controlling.

Rabbit

Probiotic has been reported to improve growth, feed utilisation and reduced mortality by maintaining microbial balance of stomach and intestine of rabbit. Very limited work has been done in India (Das *et al.,* 2001a and 2001b) and abroad also (Hollister *et al.*, 1990; Lacza *et al.*, 1990; Nievas, 1990; Blas *et al.*, 1991; Abdel, 1995; Gippert *et al.,* 1996) on the abovesaid species.

It has been found that probiotics fortification of feed improved microbial balance in the gastrointestinal tract of rabbit which resulted in better digestion and absorption of nutrients which were responsible for accelerated growth, better feed utilisation and lower morbidity and mortality in rabbit. Further, post-weaning mortality had also been observed to be reduced in probiotic fed rabbits.

Sheep/Goat

Productivity of sheep and goats in arid and semi-arid regions of India is low owing to their overstocking, depleted and poor quality pasturelands. The owners of sheep/goats do not offer any supplemental feeding to their flocks which adversely affects growth, production and reproduction performance. Probiotic preparations have a beneficial influence on animal's body weight gains and productive performance through balanced nutrition. Recently, Rohilla *et al.* (2009) found that probiotics feeding had a highly significant ($P<0.05$) effect on growth rate of Marwari kids and lactating goats as compared to control group. The kids offered probiotics attained average daily gain of 74.25 g/d as compared to control group and net returns/kid were highest from this group (Rs 1201) for the experimental period. Similarly, the goats receiving probiotics produced 45% more milk in the lactation period as compared to those goats reared under natural grazing, resulting net returns of Rs 1164/goat. However, milk composition and physiological parameters were not affected significantly. Hence, probiotic feeding to Marwari kids (for higher growth) and goats (for more milk production) has been found significantly beneficial and economical as well.

A true to Marwari breeding Buck at RRS Pali

Marwari kids weaned for experimental trial

Experimental group being offered supplemental feeding

Plate 2.

Experimental Marwari goat flock at CAZRI, RRS-Pali

Ovine rearing is economically viable in arid region

Plate 3.

Conclusions and Suggestions

Biotechnology is already widely used in animal production and there are numerous other potential applications, which can further be envisaged for improving the performance of animals through better nutrition. Probiotics or immune supplements can inhibit pathogenic microorganisms or make the animals more resistant to them.

Future areas of research that may help in the evaluation of prebiotics as potential ingredients for functional foods aimed at enhancing calcium bioavailability and protecting against osteoporosis. There is tremendous scope of probiotics in alternative use of antibiotics as growth promoters for monogastric animals for the improvement of animal performance through an improved gastro-intestinal health. To recognise the potential advantages of probiotics use we need to understand how bacterial cells function in the gut ecosystem and how further it impacts the host in question? Modulation of the bifido bacterium population in the gut, prebiotics are better candidates than probiotics, since inulin increased bifido bacterial numbers without changing the species composition, whilst the probiotics had almost no effect. Major improvements are needed in labelling and quality assurance procedures for probiotic compounds. In addition, well-planned and controlled clinical studies are necessary to delineate fully the potential for probiotic compounds. Sex related differences might be important in uptake and utilisation of energy and other nutrients and in response to anti-nutritional factors feed enzymes, prebiotics, probiotics, other feed additive and vaccinations against gut pathogens.

Bacterial species that have traditionally been regarded as safe are used in probitics, the mainly used strains are lactic acid bacteria and bifidobacteria that inhibit the pathogenic microorganism in the intestinal tract.

REFERENCES

Abdel-Sanee-AM. 1995. Using some antibiotic and probiotic for alleviating heat stress on growing does in Egypt. *World Rabbit Sci.* 3(3): 107-111.

Akman-SA and Yagci-RV. 2002. Nutritional impact of pre and probiotics as protective gastrointestinal organisms. *Annual Review of Nutr.* 22:107-138.

Apan-TZ. 2000. The probiotics and prebiotics in microbiology. Infeksiyon-Dergisi. *Turkish J. of Infection.* 14(4) : 579-583.

Baynes-P, Varley-M, Verley-MA (ed) and Wiseman-J. 2001. Gut health: Practical considerations. *British Society of Anim. Sci.* UK. 249-257.

Blas, CD, Garcia, J. and Alday, S. 1991. Effect of dietary inclusion of probiotic on performance of growing rabbit. *J. applied Rabbit Res.* 14(3): 148-150.

Cashman-K and Tannock-GW. 2002. Probiotics and calcium bioavailability. Probiotics and prebiotics: where are we going? 149-174. *Caister Academic Press*, Wymondham, UK.

Christopher, KJ. 1987. Probiotics are useful to animals and poultry. *Farmer and Parliament*. April, 1987: 19-20.

Das, SK, Das, A and Naskar, S. 2001a. Growth and feed utilisation of broiler rabbit on probiotic supplemented ration. *Indian J. Anim. Prod. Magmt.* 17: 8-9.

Das, SK, Das, A., Giri, SC, Naskar, S and Barman, K. 2001b. Effect of probiotic on performance of crossbred rabbits. *Indian J. Anim. Nutr.* (In Press).

Gandh, AB and Nagarathanam. 1990. Probiotics in Veterinary use. *Poultry Guide*. 27 : 44 -47.

Giardini-A, Salvadori-BB, Rottigni-C, Vigezzi-P, Pedron-O and Renon-P. 2001. Probiotics with lactic acid bacteria. *Rivista-di-Avicoltura*. 70(6): 14-28.

Gippert, T., Bersenyl, A., Szabo, L. and Farkas, Z.S. 1996. Development of novel feed concentrate supplemented with Salinomycin and Lacto-Saee for growing rabbit nutrition in scale farms. *6th World Rabbit Congress*. Tolouse, July 9-12. 1:189-190.

Hollister, AG, Chuke, PR, Robinson, KL and Pottorr, NM. 1990. Effect of dietary probiotics and acidifiers on performance of weanling rabbits. *J. applied Rabbit Res.* 13(1): 6-9.

Hughes-RJ and Corbett-J. 2001. Variation in the digestive capacity of the broiler chicken. *Recent Advances in Ani. Nutr. in Australia*. 13:153-161.

Immerseel-F-Van, Cauwerts, K, Devriese, LA, Haesebrouck-F, Ducatelle-R and Van-Immerseel-F. 2002. Feed additives to control Salmonella in poultry. *World's Poultry Sci. J.* 58(4): 501-513.

Karput, JM and Pudenko, LL. 1996. Bacterial preparations in the prevention of dyspepsia in piglets. *Vestai-Akademi-Agrarnykh-Nauuk- Belarusi*. 1: 73-75.

Lacza,-szabo.S, Gippert, T, Huller, I and Virag, G. 1990. Use of streptococcus Facium NM-74 in the feeding of meat rabbits. *Cunniculture*. 96 : 263-266.

Letellier-A, Messier-S, Lessard-L and Quessy, S. 2000. Assessment of various treatments to reduce carriage of *Salmonella* in swine. *Canadian J. Vet. Res*. 64 (1): 27-31.

Mandal-AB, Elangovan-AV and Verma-SVS. 2001. Interaction of nutrient and infection in poultry- a review. *Anim. Nutr. and Feed Tech*. 1(1):19-38.

Mairoka-A, Santin-E, Suguta-SM, Almeida-JG and Macari-M. 2001. Utilisation of probiotics, prebiotics or symbiotics in broiler chicken diets. *Revista-Brasileira-de-Ciencia-Avicola*. 3(1): 75-82.

Moro-GE, Warm-A, Arslanyoglu-S, Miniello-V and Fiocchi-A. 2002. Management of bovine protein allergy: New perspectives and nutritional aspects. *Annals of Allergy-Asthma and Immunology*. 89 (6-Suppl) : 91-96.

Naskar, S, Das, A and Das, SK. 2000. Effect of probiotic on growth performance of cross breed grower pigs. *Indian J. Anim. Prod. Mgmt*. 16 (4) : 142-143.

Nievas-Delgado. D.B. 1990. Effect of the amount of fibre and probiotic in the fattening of rabbits. Rebista Unellex de Ciencia Technologia Production Agricola. 10 (1-2) : 41-55.

Pradhan, RN, Sahoo, G, Mihra, PK, Babu,LK, Mishra, SC and Mohapatra, LM. 1998. Role of probiotics on performance of broiler chicks. *Indian J. Anim. Prod. Mgmt*. 14(2): 80-83.

Rekiel, A. and Wiecek, J. 1996. Influence of biogen and microferm for preparations on growth and health of piglets. *Medycyna-Weterynaryjna*. 52(3) : 187-190.

Rohilla PP, Patil NV and Bohra HC. 2009. Effects of probiotics and nutrimix feeding on kids and goats. *Indian Vet. J*. 86 : 640-641.

Spieler, A and Manner, K. 1996. Lactic acid bacteria as probiotic feed additives. *Kraftfurtter*. 5: 200-207.

Tandani, A. and Terreni, M. 1996. *Bacillus cereus* in diets for pigs. *Rivista-di-diswinewltara*. 37(12): 27-31.

Thapliyal, DC. 1993. Probiotics as animal growth promoters. *Indian Farmers Digest*. 26(12): 24-25.

Titelbaum-Je and Walker-WA. 2002. Nutritional impact of pre and probiotics as protective gastrointestinal organisms. *Annual Review of Nutr*. 22: 107-138.

Vassalo, M, Fialho, ET, Oliveira, AIG, Teiseria, AS, Bertechini, AG and Deoliveira, AIG. 1997. Probiotics for piglets from 10 to 30 kg live weight. *Rewnta-de- Socicdade-brasileira-Zootecnica*. 26(1): 131-138.

Wenk-C. 2003. Herbs and botanicals as feed additives in monogastric animals. *Asian Australasian J. Ani. Sci*. 16(2) : 282-289.

Zimmermaan, B., Bauer, E. and Mosenthin, R. 2001. Pro- and prebiotics in pig nutrition : potential modulators of gut health. *J. Anim. and Feed Sci*. 10(1) : 47-56.

23

Improvement of Sheep Productivity in Arid Region

S.A. Karim and A.K. Shinde

Central Sheep and Wool Research Institute, Avikanagar 304 501

Introduction

Sheep husbandry plays an important role in livelihood and economic sustenance of landless, marginal and small farmers in semiarid and arid region of the country. Moreover, sheep rearing is a livelihood security for certain communities in dominant sheep producing region of the country and traditional occupation of poor farmers. The sheep rearing ensures self or family employment and acts as a financial support at the time of distress like drought and famine, which is common in arid region. They act as movable asset of high liquidity and a source of household nutritional security. Sheep rearing provides meat, wool, milk and manure as major source of income and nutritional support to the family. Most of the sheep breeds found in arid region produce carpet wool which is utilized in number of rural based cottage industries, woollen mills and carpet industries. In arid region of the Rajasthan, the demand of mutton is relatively lower and majority of surplus male lambs and spent ewes are sold for mutton catering to Delhi, Jammu and Kashmir and Southern states. Milk is utilized for the family consumption as fluid milk as well as in products prepared and in some part of arid region sold through cooperative societies and farmers are getting remunerative price for milk. Sheep manure, a slow release rich source of organic carbon and nitrogen has long lasting effect on application in agriculture and also serves as an important source of earning in the hot arid region and fuel in cold arid region. The arid region is rich depository of sheep population and genetic resources and endowed with 6 well-known breeds of sheep which and many of them produce excellent lustrous wool suitable for carpet manufacturing.

Sheep population in the country is 64.2 million, ranking third in the world. Sheep husbandry contributes 243 m kg mutton, 45.5 m kg wool, 56.3 m kg skin and 190 m kg manure[1]. The arid region is host to 12.11 million sheep and important among them are Magra, Marwari, Chokla, Jaiselmeri, Nali and Sonadi breeds. The majority of sheep breeds produce ideal carpet wool in the region. As per recent figures 22.77 million kg of wool is produced in the region, which constituted 44.90% of total wool production in the country[2]. The study carried out in Rajasthan showed that major expenditure in sheep rearing is feed and fodder (56.08%), labour (9.74%) veterinary health care (28.52%) and interest repayment (5.65%). The expenditure on account of feed and fodder now is nearly 45%, which was not a major component of expenditure during earlier days. Gross return harvested from animal sale (81.36%), wool (8.65%), milk (3.26%) and manure (6.72%). Sale of young and spent sheep contribute major source of income to farmers[3].

Over the period, sheep husbandry in arid region has undergone sea changes mainly due to shrinkage of grazing land and poor wool price in the market. Sheep during early sixties and seventies were maintained under migration and nomadic life style. With advent of other source of livelihood, the preference of younger generation for sheep husbandry is declining as a source of income and employment. Moreover the grazing available for sheep over the period has declined, not only in term of area under CPR but also in term of yield and quality, which has forced them to leave the traditional occupation of sheep rearing and adopt some other occupation. Sheep are mainly raised for wool in arid region however the price of wool is declining sizably in domestic and international market: the trend is evident not only in arid region but also all over the world. The major problems faced by sheep farmers in arid region are: progressive shrinkage of grazing resources, climate change reflected in low precipitation, feed and water scarcity, frequent drought and famine, long distance migration, poor *en route* shelter, early disposal of lambs, poor wool price, distress sale of animals etc. Sheep are source of livelihood and economic sustenance of poor farmers in harsh climatic zones of the country requiring attention with respect to their productivity and sustainability. Moreover, demand of meat, milk and fibre is increasing progressively in the country requiring strategies for improving production through better feeding, breeding and health coverage. Further there is need to address the sheep production under the present scenario of land utilization, climate change and feed resource scarcity and paradigm shift in market trends for wool and meat.

Achievements/advances in sheep husbandry

The productivity of Indian sheep is lower in comparison to western countries attributed mainly to scarcity of feed resources, unfavourable climatic conditions and traditional management practices. Population of many well-known breeds of sheep in arid region is declining and the breed character is diluting because of intermixing of breeds and indiscriminate breeding on economic compulsion. Attempts made to conserve the breeds *in situ* has not largely succeeded due to faulty implementation and social factors However NBAGR and other institutions are making attempts for conservation of threatened breeds. Sheep depend on grazing resources as major source of feed and over the period the grazing lands and biomass yield of lands are deteriorating hence an alternative system of grazing plus supplementation has been developed by the CSWRI and advocated as a development strategy in sheep production. Moreover CSWRI, IGFRI, CAZRI and other ICAR institutions and SAU are propagating rehabilitation of community land by establishing perennial grass pasture and silvipasture. The diseases of parasitic, bacterial, viral and fungal origin cause economic losses, systematic disease control programme is suggested through state departments to reduce morbidity and mortality thus net economic losses to farmers. In arid region sheep are exposed to drought and famine conditions often compelling many flocks to migrate to neighbouring states. Complete Feed Blocks of roughage and concentrate in variable proportions are developed by the CSWRI meeting requirements of various physiol-ogical stages and production levels for feeding during scarcity period. Moreover several newer feed resources like Opuntia, tree pods and leaves are being explored that can be fed to animals during drought period. Some of the advances made in the enhancing mutton, wool and milk yield potential of sheep of arid and semiarid regions are detailed below:

Mutton Production

The mutton is a primary produce of sheep rearing accounting for 70-75% of farmers earning. About 32-35% of sheep are slaughtered annually however the slaughter rate is relatively lower compared to neighbouring countries ascribed mainly to lower lambing rate (65-70%) and prolificacy of sheep under field conditions. Hence higher prolificacy with twins and triplets are an essential trait for successful mutton production programs. Earlier male lamb surplus to the breeding requirement used to be marketed for slaughter around 9-12 months of age weighing about 22-23 kg providing average carcass weight of 10 kg. However with the evident climate change in the region and inability of farmers to maintain these lambs in the flock has led to visible change in the *in vogue* production system wherein such lambs are being disposed off at 3-4 months of age weighing about 16- 17 kg providing carcass of 8 kg. In order to meet the requirement of higher pre weaning growth in lambs, the farmers introduced Patanwadi sires with proven higher milk yield in the flock. It is being envisaged in the development programs to involve entrepreneurs/progressive farmers in the trade to purchase these weaner lambs from the sheep farmers and rear them up to finishing stage and harvest 30- 33 kg at 6- 7 months of age under protocol of intensive feeding (R:C, 40:60) or grazing with *ad libitum* supplementation. Three decades of research at CSWRI, Avikanagar showed that the growth rate of native sheep could be improved to the tune of 300% from 40- 50 g to 150-170 g with 16-18% feed conversion efficiency on complete feed under intensive system[4] or grazing with supplementation. Native Malpura lambs attained 33.00 and 27.00 kg finishing weight at 6 months of age with 50- 55% dressing yield under intensive and semi-intensive feeding systems. The developed technology on mutton production has been successfully demonstrated on farmer's field but it is not largely adopted due to poor economic background of the sheep farmers and lack of monetary support from financial institutions in term of feed input. Under organized intensive feeding system and in the present cost of feed input/kg live weight gain and per kg dressed meat yield, it comes to Rs. 28.00 and Rs. 56.00, respectively while the selling price of meat is Rs.150-200/kg in different parts of the country[5]. The simple arithmetic of input and output ratios indicates the scope of profitability in the mutton production venture. The slaughterhouse by-products viz. edible offal, head and hooves, skin and viscera fetched on an average Rs. 250/animal. It is generally accepted that the return from by-products are the net profit of the butcher while the return from the dressed carcass is the purchase price of the animals, which in turn, is shared by the primary producers and the middlemen.

About 40- 50% spent category of sheep after completing their productive life are put to slaughter for meat purpose. The carcass emerging from spent sheep have dressing yield ranging from 38-40% with higher proportion of bone and lower proportion of fat and muscle. Intensive feeding of the spent animals for 45-90 days before slaughter increased the finishing weight by 7-10 kg and carcass fat to 12-15%, rendering the carcass to acceptable level for meat purpose. Although under protocol of pre-slaughter feeding of spent sheep, the cost of feed input/kg live weight gain is relatively higher but the production traits, meat yield and quality are sizably improved. The cost feed can be minimized utilizing locally available feed ingredients and least cost feed formulations[6].

The sheep rearing is predominant in arid agro-climatic condition, which is essentially deficient in feed and fodder resource while feed resource rich regions are devoid of sizable sheep population. It is, therefore suggested that the weaner lambs could be procured from arid region and feed ingredient from feed resource rich region to establish lamb rearing centres near some exit port or metropolis and the finisher lambs can be marketed effectively for profitable mutton production. Entrepreneur may be involved in the venture right from procurement to marketing chains. Moreover, in intensive mutton production program, the cost of feed input alone accounts for more than 70% of total expenditure hence the sheep farmers who are used to virtually zero input on feed, are generally reluctant to adopt high feed input for mutton production. In active phase of growth, the cost of feed input/kg live weight gain average to Rs. 30/kg gain. Accordingly the cost of feed input/kg meat production comes to Rs. 60 which selling @ Rs.170-300/kg in different parts of the country indicate the margin of profitability for the operators and commercial viability of the venture. Although there is considerable margin of profit in lamb rearing for meat production, the business houses and entrepreneurs avoid investment in the enterprise due to social taboos. The weaker segments of society involved in the sheep husbandry neither have financial resources at their disposal to establish organized mutton production system nor literally educated to develop linkage with collateral agencies involved in fattening and trading. Financial institution may provide soft loans to the farmers for ensuring economic and nutritional security of poor farmers engaged in sheep farming.

Wool Production

Wool is unique product of sheep produce having wide diversity ranging from coarse wool to apparel fine wool and hairy fleece in different parts of the country. Indian wool is used for making hand-knotted carpets, druggets, hosiery items etc. for domestic and export market. The scope for apparel wool production in the country is marginal and restricted sub temperate and temperate locations while carpet wool can be produced in semiarid and arid region meeting the demand of carpet industries. The fact remains that domestic production of carpet wool meets 40% requirement of carpet industry hence import from wool producing will continue. However situation in demand and supply of wool clearly indicate the scope of manipulations for improving wool production. Sheep husbandry is dominated in western parts of Rajasthan. Marwari, Jaiselmeri, Chokla, Magra, Nali etc. are important wool breeds of arid region. There are about 14.40 million sheep population contributing more than 56% of total wool produced in the country and most of the wool is suitable for carpet manufacturing. About 28.19 million kg wool is produced in western Rajasthan with average per sheep yield of 1.95 kg with fiber traits of fineness, staple length and medullation of 30.45μ, 4-5 cm and 30.80% respectively[7].

Wool markets in Rajasthan are not efficient in transmitting the prices. The decomposition analysis of wool prices for the period of 1995-96 to 2005-06 indicated that the wool price is mainly affected by trend, cyclical and seasonal fluctuations. The seasonal index indicated that the price falls during the winter period. Low price of wool, collusion of traders in the market and low market infrastructure and high market fee and reductions were identified as the major constraints in wool marketing. Coarse wool has little utility in textile and various other products can be developed from blending of coarse wool with fine wool and other fibres for value addition.

The competition of wool with man made fibre is main factor threatening the growth of woollen sector in the country besides increasing import of carpet and apparel wool from the other countries. Wool price in the country is very low and not increasing in relation to other animals produce. The provision of support price for wool may be implemented for providing better price to producers and also for safeguarding their production cost.

Milk Production

Sheep milk is of limited importance except in areas of Rajasthan and Gujarat where surplus milk is utilized by shepherds' families. At present sheep milk is not marketed in the country probably due to smaller quantity of milk produced or lack of awareness among consumers about its utility as milk in human beings. The survey carried out in Rajasthan indicated that on average 4-5 kg of milk is available to farmers from a flock of 100 sheep. Major portion of it, is consumed in houses as fluid milk. In the country, sheep is not considered as dairy animals as in Middle East and Eastern European countries of the world. Malpura and Patanwadi are important sheep breeds of Rajasthan and Gujarat respectively known for higher milk yield in the country. Malpura sheep yield 515.7 g of milk daily and Patanwadi sheep of Kathiawad region of Gujarat yield 783.7 g of milk daily in a lactation period of 90 days [8].

On an average, sheep milk of Indian breeds contained 16.30% total solid, 5.30% fat, 5.50% protein, 4.60% lactose and 0.90% ash. Sheep milk is highly nutritious, richer than cow milk in vitamins A, B, and E, calcium, phosphorus, potassium, and magnesium. It contains higher concentration of short and medium chain fatty acids, which are beneficial to health of human being. Short-chain fatty acid is easily digested and has little effect on cholesterol. Sheep milk has more conjugated linoleic acid (CLA) - a cancer-fighting agent, than the milk from goats and cattle. Sheep milk contains smaller fat globules than cow milk, making them easily digested. Sheep milk in the country shall be popularised and explored for marketing as a regular source of income to the farmers.

Gaps and Challenges Ahead in Sheep Husbandry

Under present scenario of climate change affecting sheep population and grazing land in arid region of the country, drastic changes in term of population and production is expected. Sheep are basically raised under extensive range management moreover in arid region more than 90% of flocks are managed under migration of varying nature extending from 3-4 month of short migration to round the year migration of permanent nature. The water and feed scarcity in arid region under climate change will aggravate the situation and make sheep rearers to either maintain animals on migration or leave the occupation. The situation will become severe during drought and famine in arid region compelling farmers to take their animals on migration or resort to distress sale. There is an imperative need for developing policy for feed and fodder in arid zone. The grazing policy for the region may be developed and migration policy may also be strengthened with supports from forest, revenue and animal husbandry departments. Prolificacy is an essential trait for success full sheep production program. A three breed cross involving native Malpura, Marwari and Magra sheep with introgression of prolificay trait from Garole and milk yield support from Patanwadi will meet the requirement of higher mutton and wool production. Like wise to address low reproductive efficiency Cryo-preservation of semen and artificial insemination is basic requirement for rapid improvement of flock productivity by increasing the population of elite animals in the flocks. The technique of frozen semen of ram is perfected but the fertility of frozen thawed ram semen insemination is low moreover the use of trans-cervical artificial insemination using frozen semen still require perfection for desirable fertility. Wool is main produce of sheep and price of wool over the period not increasing as in other commodities. Sheep breeds found in arid region produce excellent carpet wool, but farmers are not getting remunerative price for their produce. Wool price has to be insulated against market variability to safe guard the interest of farmers. Early disposal of male lambs for slaughter is prevalent practice in the region because of feed scarcity and fear of mortality losses, the provision of integrated programme of supplementary feeding and prophylactic and curative health cover need to be implemented.

Research and Strategies for Enhancing Mutton, Wool and Milk

The enhancement of productivity of sheep in term of mutton, wool and milk to meet the growing demand of ever increasing human population is one of the major concerns of sheep husbandry in the country. In present scenario of shrinkage of grazing land, the improvement of productivity per sheep is an important aspect. Some of the technologies developed for enhancing sheep production of field flocks are detailed below:

The demand of mutton is increased exponentially due to increasing economic status and health consciousness and general awareness of consumers. In the recent past, looking to higher earning from mutton than wool, preference of farmers has changed toward heavy body weights breeds for mutton production than wool breeds. Accordingly to meet the requirement a three breed cross with Malpura, Marwari and Magra is being developed incorporating prolificacy traits in non prolific flock from Garole and a terminal cross with Patanwadi for higher milk yield support. Synchronization of oestrus and reproductive manipulation with over all strategy of harvesting three lamb crops in two years will further enhance mutton and wool production.

Prolificacy traits and lambing frequency can increase the number of lambs for mutton production. Garole and Kendrapada prolific sheep breeds from Sundervan areas of West Bengal and costal region of Orissa respectively are well-known prolific sheep breeds and can be used for introducing prolific traits in non-prolific sheep breeds. The study carried out at CSWRI, Avikanagar indicated that cross breeding of prolific Garole with non-prolific Malpura breed increased prolificacy from 1.05 to 1.60 in Malpura x Garole crossbred sheep. Average lambing percent improved from 91.75 in Malpura to 148.21 in GM sheep. The low twining percent of 4.71 in Malpura improved to 51.10 in GM sheep[9]. Mutton type Indian sheep breeds having higher feed efficiency, growth rate and dressing yield can be brought under the program for enhancing mutton production by introgression of *Fec b* gene from Garole or Kendrapada sheep.

At present under prevailing breeding practices farmers take around one lamb per sheep per year, there is every possibility of harvesting three lambs in two years under accelerated lambing programme. The introgression of prolific gene and increasing of lambing frequency would increase mutton production through more number of lambs available for slaughter.

Lambs attained body weights of 8-10 kg at weaning age (3 months) and 16-18 kg at 6 months of age in the field flocks. Intensive feeding protocol in the lamb feeding has ability to increase the pre weaning body weight to 18.00 kg and post weaning finishing weight to 33.00 kg in Malpura lambs with feed efficiency of 17-18% which is commercially viable. Feed cost per kg weight gain and mutton production comes to Rs. 28.00 and 56.00 under intensive and Rs 21.00 and 42.00 under semi intensive system respectively while mutton is sold at the rate of Rs 150-250 in different parts of the country. The commercial viability of the venture is evident from the data however the major limitation in commercialization of trade is poor socio economic status of the farmer and inhibition of entrepreneurs in investing on any thing dealing with meat.

Mineral deficiency in sheep maintained on community grazing lands is common problem in tropical country especially in arid and semiarid topography. Mineral deficiency severely affected mutton, wool and milk production. CSWRI, Avikanagar developed area specific mineral mixture for semiarid region of Rajasthan and supplementation of it in sheep flocks regularly improved the milk yields by 15-20%, wool yield by 10-15%, bringing anoestrus sheep into oestrus within 21-30 days of supplementation and eases the problem of skin keratinisation[10]. The mineral mixture supplementation in sheep flocks shall be popularised for enhancing production.

During lean season sheep generally ingest soil along with forage on sparse vegetation cover in desert regions due to their close grazing habits. Soil ingestion increased excretion of endogenous nitrogen per unit of faecal organic matter besides damage to gut mucous lining, rapid wearing of incisor teeth and brushing of dental pad. Moreover soil ingestion increase during drought and famine situation because of poor vegetation cover. The study conducted at CSWRI indicated that sheep ingested soil up to 39.10, 15.62 and 46.41% of total dry matter intake during monsoon, winter and summer and severely affected nutrient digestion and also causing death in some cases[11]. Some kind of supplementation in form of grass hay or other feeds may be provided to improve the situation.

Feed scarcity during critical season compels the farmers to sell off their young stocks at an early age. Complete Feed Blocks of roughage and concentrate mixture has been developed at CSWRI to facilitate storage, transportation and fortification of low-grade roughage for improving nutrient intake, utilization and avoiding feed selectivity. The feed blocks could serve as source of sustenance feed for livestock species in famine prone zone of arid region[12]. Moreover feed blocks can be fed as complete feed for enhancing mutton production especially in feed deficient zones of the country.

The demand of lambs for slaughter fluctuate widely during the year, demand increased during Muslim festivals and during winter season. The concept of rearing lambs as per market demand for harvesting better price is possible by manipulating reproduction. Impregnated sponges are developed in the CSWRI, Avikanagar for oestrus induction and synchronization in the sheep flock[13]. These sponges are in demand among the sheep breeders for inducing oestrus in sheep as well as for synchronizing the lamb crop to match the demand in the market. There is need to develop state of art of semen collection, semen processing, semen freezing semen storage and semen utilization facilities meeting stringent sanitary and phyto-sanitary requirements for ensuring the availability of semen of elite rams for AI.

Under prevailing sheep rearing practices under field condition, mortality and morbidity in animals cause economic losses to the farmers and severely affect their livelihood and economic sustenance. Sheep Flock Health Management Calendar has been developed based on the disease data information system. A low cost package of health practices, which involved vaccination against PPR, ET and Sheep pox, one strategic drenching for de-worming and one dipping has been recommended. Sheep are drenched 3- 4 times in year with anthelmintic under traditional rearing in the region. The implementation of calendar in field flocks reduced the possibility of any outbreak of diseases and mortality rate to >5 per cent per annum. This increased expenditure and risk of resistance against medicine. The Institute has developed strategic drench programme, which has saved the expenditure up to 63% on anthelmintic without any decline of production[14]. Application of Flock Health Technology costing Rs. 3000/100 sheep/year will save 8 animals costing Rs. 12000 from evident death providing net saving of Rs. 9000 which has to be demonstrated to sheep farmers in economic terms.

Sheep are raised under extensive range management, however over the period grazing land is shrinking, an alternative system of grazing and strategic concentrate supplementation for mutton and wool production is popularised in sheep dominating regions. In sheep rearing, more than 70% earning come from sale of meat. Marketing of finisher lambs and spent animals as separate identity will provide better return to farmers. The meat from finisher lambs can be marketed as different cuts with variable price for choice cuts while those from spent animals can be converted in to heat and eat finished products. India has rich diversity of culinary specialties, these recipes have been scientifically documented for their commercial preparation, packaging, shelf life enhancement, keeping quality evaluation and marketing. There is good scope of increasing export of meat and processed mutton products to Middle East and South East Asian countries hence production of desirable body weights and carcass quality including disease free animals are to be strengthened.

Sheep breeds of arid region produced wool suitable for carpet manufacturing, Magra, Chokla, Marwari sheep wool are mostly used in carpet manufacturing and employment opportunity for peoples in the region. Camel, equine and goat hair can be blended with wool to produce customer attractive furnishing fabrics for medium to rich end users. Low cost camel hair blankets are prepared (approximate cost Rs. 15/kg) by blending with wool (approx. cost Rs. 60/kg) for low and poor income users. Indigenous technology of wool utilization developed by the CSWRI for Khadi, felt industry, carpet industry and local artisans are helpful in increasing value addition and employment opportunity in rural areas.

At present, major portion of sheep milk is consumed as fluid by the shepherd community and very little is being sold in the market or converted into milk products. In Gujarat, sheep farmers convert some portion of surplus milk into ghee and dahi for their family consumptions. Mozzarella cheese made from sheep milk made is famous in western countries. Sheep milk in compared to goat or cow milk has higher solid content, as a result, more cheese can be produced from sheep milk. Sheep milk yields 18 to 25% cheese, whereas cow and goat milk only yield only 9 to 10%. Sheep produce less milk than goats and much less than cows hence sheep milk is sold at a significantly higher price per kg, almost four times the price of cow milk in western countries. Sheep milk may be popularised in product preparation like cheese, this will generate employment opportunity in the rural areas.

Drought and famine are common in arid regions of the country causing severe economic loss to farmers by reduction in production, distress sale and disposal of prime animals and death. It also suppresses breeding, fertility, pregnancy and lactation in animals. Poor nutrition during drought cause poor body condition and lower resistance in animals succumb to mortality and death. Drought condition cause seminal degeneration, resulting in poor quality semen, spermatogenesis and normal spermatozoa. Heat and nutritional stresses reduce sexual drive and breeding activity in rams during drought. Water deprivation is common problem in arid region and situation become still grave during drought period. Moreover, saline water is another problem in desert severely affecting the health, reproduction and production. Sheep are exposed to saline water for more than 8 months in desert and tolerate mild concentration of TSS in water however when concentration is more than 4.5 g/litre their reproduction and production is affected beside diarrhoea, emaciation, depression of oestrus cycle and early embryonic losses. Drinking water in desert also contained higher proportion of fluorine and nitrate and higher concentration of fluorine and nitrate 1.5 g/l adversely affected the production and reproduction[15]

Drought mitigation plan shall be developed to protect the economic losses to farmers from loss of production and death. The preparedness and mitigation plan should include provision of feed and fodder on nominal rates, drinking water, protection of distress sale of animals, *en route* support during migration to neighbouring states, minimum entry charges in other states, treatment of animals, compensation for death losses etc.

In arid region, sheep husbandry is back bone of rural economy and source of livelihood and economic sustenance for the people, moreover productivity of sheep in the region under prevailing rearing system is economical but it would not be possible to meet the growing demand due to climate change and shrinkage of grazing resources. There is need to integrate the program of breed improvement, enhancing reproductive efficiency, supplementary feeding, prophylaxis and curative measures and marketing infrastructure and mutton and wool processing and value addition to increase per sheep productivity and more return to farmers from sheep rearing.

REFERENCES

1. Food and Agriculture Organisation, 2007. Livestock Statistics, FAO, Rome
2. Shinde, A.K. 2009. Wool yield and quality under different production systems In: Wool Technology- Innovation in wool production and technologies for value addition (Ed S.A. Karim, D.B. Shakyawar and Anil Joshi), Agrotech Publishing Academy Udaipur
3. Suresh, A., Gupta, D.C. and Mann, J.S. 2008. Socio-economic status of sheep farmers and marketing structure for sheep products in India with particular emphasis on semi-arid regions of Rajasthan. In: Small Ruminant Production in India -Strategies for enhancing (Eds Karim, S.A., Tripathi, M.K., Chaturvedi, O.H. and Shinde, A.K.) Satish Serial Publishing House, New Delhi, pp 503-518.
4. Karim, S.A. Santra, A and Singh, V.K. 2004. Lamb feeding for mutton production, Bulletin, CSWRI, Avikanagar
5. Karim, S.A. 2008. Perspectives of small ruminant production for augmenting rural prosperity. In: Small Ruminant Production in India - Strategies for enhancing (Eds Karim, S.A., Tripathi, M.K., Chatur-

vedi, O.H. and Shinde, A.K.) Satish Serial Publishing House, New Delhi, pp 1-17.

6. Bhatt, R. S., Tripathi, M. K., Soren, N.M., Jain, P.K. and Karim, S.A. 2009 Exploitation of genetic potential of growth of native animals by challenge feeding. Annual Report, CSWRI, Avikanagar

7. Karim, S.A., Shakyawar, D.B. and Joshi, Anil 2009. Wool Technology: Innovation in wool production and technologies for value addition. Agrotech Publishing Academy Udaipur

8. Naqvi, S.M.K., Gulyani, R., Saha, S., Singh, Sajjan, Joshi, Anil, Maurya, V.P. and Kumar, Davendra. 2009. Milk yield of Malpura and Patanwadi sheep, Annual Report CSWRI, Avikanagar

9. Mishra, A.K., Arora, A.L., Kumar, Sushil, Kumar, Satish and Singh, V.K. 2007. Improving productivity of Malpura breed by crossbreeding with prolific Garole sheep in India. *Small Ruminant Research.* 70: 159-164.

10. Gowda, N.K.S., Pal, T., Gupta, Rajan, Prasad, C.S. and Sampath, K.T. 2009. Feed resources and feeding practices in different agro-climatic zone of India. National Institute of Animal Nutrition and Physiology, Bangalore pp 469

11. Shinde A.K., Sankhyan, S. K. Singh N.P. and Verma D.L. 2005. Soil ingestion and its effect on dry matter intake and digestibility in sheep grazing on native ranges of semiarid region of Rajasthan. *Indian Journal of Small Ruminant.* 11(1): 37-42.

12. Jakhmola, R.C. 2008. Interventions to sustain small ruminant productivity during drought conditions. In: Small Ruminant Production in India - Strategies for enhancing (Eds Karim, S.A., Tripathi, M.K., Chaturvedi, O.H. and Shinde, A.K.) Satish Serial Publishing House, New Delhi, pp 193-209.

13. Naqvi, S.M.K., Joshi, A., Das, G.K. and Mittal, J.P. 2001. Development and application of ovine reproductive technologies: An Indian experience. Review *Small Ruminant Research* 39: 199-208.

14. Swarnkar, C.P. Singh, D. Krishna Lal and Khan, F.A. 2008. Epidemiology and management of gastrointestinal parasites of sheep flocks in Rajasthan. Bulletin CSWRI. Avikanagar, pp. 145.

15. Mittal, J.P. 1994. Tolerance of Indian desert sheep for prolonged intake of fluoride in drinking water, Seventh AAAP Animal Science Congress, Indonesia, pp 142.

24

Drought adapted breeds of livestock

S.K. Kaushish

Central Arid Zone Research Institute, Jodhpur - 342 003

1. Introduction

The Indian arid zone covers an area of about 32 million ha. The large bulk of the area lies in the arid zone of Rajasthan. Only 38% of the Rajasthan population lives in the 61% area, which forms a part of Great Thar Desert. Livestock population in India is 482.4 million while in Rajasthan it is 49.1 million. Rajasthan has the distinction of having the largest population of cattle, sheep and camel. According to 2003 livestock census, there were 10.8, 10.4, 10.0, 16.8 and 0.5 million cattle, buffalo, sheep, goat and camel, respectively in Rajasthan, while very small increase in cattle population was observed in arid Rajasthan (18.82%) and almost nil in entire Rajasthan (0.83%) as a whole during this period. District wise population in 2003 of livestock in arid regions of Rajasthan is shown in Table 4.

Despite of large number of livestock population and their production in the state, the per unit productivity of livestock is low due to decreasing grazing area, periodic droughts and erosion in good genetic materials.

Livestock rearing is an important occupation of the farmers of the arid areas particularly for augmentation of their income during drought. During 1956-97, the livestock population increased in arid Rajasthan from 13.4 million to 28.6 million. During this period there has been an increase in buffaloes, sheep, goats and camel population by 188.2, 93.2, 94.2 and 52.2%, respectively. Livestock of the arid areas is known to inherent high production potential and drought hardiness. Cows of Tharparkar, Rathi, Gir and Kankraej breeds have high milk production capacity (1,500-2,500 litres per lactation) and Nagauri, Hariana and Kankrej bullocks have reputation as excellent draught animals.

Goats are reared both for meat and milk in addition they also provide fiber and manure. One third of the total meat requirement of the country is met out from goat meat. Rajasthan state is also endowed with the best breeds of goat for both milk and meat. The important breeds of goat of Rajasthan are Marwari, Sirohi, Jakhrana and Parbatsar. Some of the goat breeds may yield up to 2 kg milk per day. Marwari, a desert goat breed can survive even on meager quantities of feed, fodder and water.

Milk has attained the status of being as the record largest agricultural commodity next only to rice in the country. At present the milk production in the country is 85.6 million metric tonnes and is expected to reach 90 million metric tonnes by 2010 AD. The western part of the state has comparatively lower population of milk animals. This area is more drought prone and has less fodder to feed the milk animals. However, the best milk breeds of cattle like Tharparkar are found in Jodhpur, Jaisalmer and parts of Pali district; Rathi in Bikaner district and Kankrej (Sanchori) in Barmer and Jalore districts of western part of the state.

Sheep and goats have been indiscriminate as the creature of desert conditions because of their close selective and voracious eating habits and to travel over long distances in search of fodder and water. They consume many weeds and bushes. The principle source of meat in the country is from sheep and goats. About 40% of these are slaughtered every year.

2. Importance of Livestock Farming

The arid region is the home tract of Tharparkar, Kankrej and Rathi breed of cattle. These breeds are recognised for their milk production and efficiency. The Nagauri bullocks are well known as draught animals. Murrah, Mehsana, Jaffarabadi and Surti breeds belonging to arid and semi arid regions are very good milk producers. Chokla, the medium fine wool breed is found in arid region. Nali, Lohi, Magra, Marwari, Jaisalmeri and Pugal, which are major carpet wool producers, also belong to this region. Sonadi and Malpura are good meat producers. Bikaneri, Jaisalmeri and Marwari camels are inhabitants of this tract. Marwari and Kadiawari horses are among the best race horses of the country.

2.1 Livestock farming in drought proofing

Due importance has not been given to livestock farming in the past. There had been more emphasis on crop farming. Agriculture is more affected by drought as compared to the livestock. During drought years agricultural production may reach as low as 10% of the normal year, whereas production may still remain more than 50 per cent under same conditions from livestock. In a year with less than 25% of average rainfall, as much as 20% of the livestock are wiped out and in a year with less than 50% rainfall the livestock population is reduced by about 10% in arid region. Livestock farming has, therefore, been recognized as an instrument of drought proofing. The major components of Drought Prone Area Programme are sheep and goat development for wool, mutton and dairy development.

Livestock farming contributes significantly to the economy of the arid region but because of management needs it involves much more intensive use of labor as compared to crop farming. Dairy enterprises involve more intensive use of labor as compared to sheep and goat rearing. On an average 5.5 man hours per household per day during monsoon are utilized in dairying and only 1 man-hour is utilized per adult cow unit. Therefore, at the present level of production with cattle and cow unit, dairy alone provides employment of 10 million man-hours per day. An equivalent amount of employment is estimated to be available from sheep, goat and camel farming. The goat dairying is affected to a lesser degree than cattle dairying during droughts particularly when the land is affected by salinity and wind erosion.

2.2. Breeding livestock for drought prone areas

The state produces about 5% fine, 60% medium and 35% coarse quality wool and as a consequence majority of fine wool for apparels is imported. With a view to improve the breed of sheep and also to increase productivity per animal large scale crossbreeding programmes were launched. The state has 3 departmental sheep breeding farms where half-breds are produced by crossbreeding native sheep with imported improved germ plasm. These half-breds are distributed among shepherds for breeding purposes. As a consequence quarter breds are produced which produce finer quality of wool as well as higher production of wool per crossbred sheep. This programme is under implementation in 12 districts of the state namely Jaipur, Ajmer, Udaipur, Bhilwara, Tonk, Chittorgarh, Banswara, Churu, Jhunjhunu, Sikar and Doongarpur.

3. Importance of Livestock in Mitigating Drought

The Livestock is a vital sector of India's agrarian economy, more particularly the rural economy. It generates potential for self-employment. The livestock rearing activity also provides subsidiary occupation to a large section of society particularly to the people living in the drought prone/rainfed areas where crop production on its own may not be very productive and engaging them fully or partially. The main assets of desert landowners are large herds of livestock. In absence of favourable condition for agriculture, livestock rearing is the only alternative source of livelihood for majority of the rural population in arid zone. Animal husbandry has been and continues to be the main stay of most of the inhabitants. The livestock to some extent absorbs the losses in crop production due to weather induced uncertainties. The contribution of the livestock sector in national economy can be assessed from the fact that the gross value of output from livestock sector alone, at current prices of 1998 was about Rs 827 billion in 1995-96 which was equal to about 26% of the value of total output from the agricultural sector. This excludes the contribution of animal draught power.

Intense heat and lack of water are two major constraints faced by the livestock in arid region, which restrict expression of their full genetic potential. The livestock may receive some care, but it may not be possible to convert the harsh environment to hospitable one, because these animals have to graze or browse in open fields on an average for 10 hr day^{-1}. It is during this time that animals are exposed to very high temperature.

3.1 Cattle

In the arid region of Rajasthan, there are some important dairy (Tharparkar, Rathi, Red Sindhi), draft (Nagauri) and dual-purpose (Kankarej) breeds of cattle. The Haryana breed of cattle is common in the arid part of Haryana and Gir breed of cattle is commonly found in arid part of Gujarat. On an average, the milk yield per lactation (305 days) of Gir, Haryana, Kankarej, Nagauri, Rathi and Tharparkar breeds has been reported to be 2500, 2000, 2000, 1000, 2500 and 2500 litres, respectively.

• ***Tharparkar***: Once common in most parts of the country, the Tharparkar pure blood is rare now. Good specimens of the Tharparkar breed of cattle are rarely seen even in Jodhpur and Jaisalmer districts once considered to be the home of this breed. A detailed survey has shown that the advantage of cross-bred cows over pure Tharparkar, are greatly disturbed by the susceptibility of the crossbred to heat, their high cost of management and health cover. Tharparkar breed of cattle is well adapted to arid environmental stresses. The heat tolerance co-efficient value (88.88 ±1.92) of free grazing animals of this breed was equal to animals of Rathi breed (89.87 ±. 2.51). During summer season, there was an increase in pulse and respiration rates. These parameters were found to be the highest in June and lowest in January. The daily water consumption per 100 kg live weight was also highest (13.27 litre) during May and lowest (5.62 litre) during December. The studies on the effect of improved management on the productivity of Tharparkar cattle showed that the productivity of Tharparkar cattle in terms of lactation milk yield, lactation length, dry period, inter-calving interval, milk fat, solids not fat (SNF) and total solids (TS) was better in cows maintained under scientific management, than those under traditional management (Patel *et al.,* 1994). The average lactation yield under improved management was 1,606 kg in a lactation of 293 days. The average peak yield was 9.67 kg in a day. The age at first calving was less than 3 years, which is similar to that of crossbreed cattle in India. Average calving interval and dry period were 364 and 76 days, respectively. No adverse effect of summer stress on the fertility was observed, as 80% cows conceived between April and August. The milk composition of Tharparkar cows was influenced by season. The highest milk fat content was found in winter season, followed by autumn, rainy and summer seasons. However, no seasonal effect was observed on milk fat under arid conditions (Patel *et al.,* 1997).

• ***Rathi***: Rathi cattle are native of Bikaner region. This breed is a milch type, brown in colour and of large size. A small herd of Rathi cattle was maintained at Regional Research Station of CAZRI for studying its productive, reproductive and adaptability performance under improved management in its native tract. The Rathi cattle attained highest (322.9 ±13.17 kg) and lowest body weight in October and August (246.31±1.90 kg), respectively. They attain puberty when they attain weight between 208 and 302 kg. During pregnancy, animals gained around 53 kg in body weight.

Animals started losing body weight after calving and it continued till 90 days after parturation (Kaushish *et al.*, 1998).

- ***Kankrej*:** Kankrej breed of cattle is commonly found in Barmer and Jalore districts of western Rajasthan. Cows of this breed have excellent milk-production capacity (1,500-2,500 litre per lactation), and bullocks are excellent draught animals. Studies were conducted at Bikaner with a herd of 16 heifers and one male calf of Kankrej breed of cattle on traditional management prevalent in rural areas. Animals were maintained on *Lasiurus sindicus*-dominated pasture at a stocking rate of 1 animal per 1.5 ha. During lean period in addition to grazing, animals were given pelleted concentrate mixture and *Lasiurus* grass hay. During 18 months, average body weight gain per animal was 67 kg. Up to the age of 4.5 years, none of the heifers exhibited signs of oestrus. The male calves also did not show any symptoms of maturity up to the age of 4-5 years in spite of their company with the heifers all the time. It may be concluded that the growing Kankrej heifers must get adequate nutrition to mature at an early age (CAZRI, 1991).

3.2 Buffaloes

The buffaloes mostly found in Rajasthan are of the Murrah and Mehsana type breeds found in the northern and southern parts of the desert, respectively. These buffaloes, on an average, yield 3000 litre of milk per lactation period of 305 days. Buffaloes like to wallow in water. Earlier it was thought that these do not adapt to hot arid climate. During the past 40 years buffalo population of Rajasthan increased from 3.0 to 7.7 million. In arid-districts of Rajasthan there were 0.722 million buffaloes in 1951 and these increased to 2.297 million in 1992. People have developed liking for buffalo milk (Kaushish *et al.* 1998).

- ***Buffaloes in rural and urban areas*:** The age at first calving in buffaloes of rural sector averaged 55 months in comparison to an average 48 months in urban sector. The service period ranged from 10 to 12 months in rural areas and 2.5 to 8 months in urban areas. The intercalving periods were 20-22 months in villages as compared with 13-20 months in Bikaner city. Although the incidences of silent oestrous/anoestrus were observed in buffaloes of both the places, these incidences were significantly higher in rural sector. Incidences of repeat breeding were as high as 50-60% in buffaloes managed in villages and as low as 10-20% in buffaloes managed in city. This study indicated that scope for rearing buffaloes in urban areas of arid region is far better than rural areas.

3.3 Sheep

India ranks fifth among the sheep rearing countries in terms of numbers but occupies tenth position in wool production. Average per head per annum wool production of our sheep is 0.88 kg but varies from 4.0-5.0 kg in some advanced countries. The important breeds of sheep in the desert region and their wool production per year are Chokla (1.54 kg), Magra (1.80 kg), Marwari (1.38 kg), Nali (2.12 kg), Pugal (1.67 kg) and Jaisalmeri (1.65 kg). On an average, 1.56 kg of wool per sheep per year is obtained in desert districts which is higher than the average wool obtained from other parts of the state (1.38 kg) and the country (0.88 kg). Sheep rearing in our country is in the hands of the poor people, who occupy very low status in the society. Low income from the poor strains of sheep, unorganised markets for the produce and poor quality pastures also contribute to loose interest in the sheep rearing. Shepherds follow age-old practices of sheep management, which not only affects the capacity to produce but also reduces the productive life span of the sheep (Kaushish, 1991, Kaushish *et al.,* 2006).

About 51% of the total sheep population of this country is found in southern plateau. Half of the sheep in this region, especially in the coastal areas, produce no wool and rest produce extremely coarse, hairy and coloured fleece. Similar is the quality of wool in the breeds of sheep in the eastern region. Except for Bennur and Nilgiri, they are hairy. These two are only carpet wool producing breeds of this region. The sheep of this area are mainly reared for meat production. Indian sheep are well known as superior carpet wool producers. The best carpet-wool producing animals are found in the northwestern arid and semi-arid parts of India. In the northwestern arid region approximately 23 million kg wool is produced annually. The total wool production in temperate Himalayan region, Deccan plateau and from eastern region is 3.0, 7.2 and 1.4 million kg, respectively. These breeds except Malpura, Sonadi, Muzaffarnagari and Jalauni, produce medium to fine carpet wool. Marwari in northwest and Bellary/Deccani in the south comprise the most popular breeds (Kaushish, 1992).

• *Physiological investigations on heat and water stresses*

Availability of drinking water being the most critical limiting factor for survival in the desert, a detailed comparative study of the physiological responses of different desert sheep breeds to imposed heat and water stress has been made. These studies have pointed out the unusual ability of the desert breed (e.g. Marwari) to maintain circulation even when faced with considerable haemo-concentration. When water intake was reduced to below 75% of the normal daily requirement, there was a steady decline in the body water stores. The desert sheep apparently relies on reduced urine and faecal water output as the means for combating water stress. This, along with their ability to derive sustenance from poor quality forages helps in maintaining animal productivity in this region. Studies conducted at CAZRI show the physiological superiority of the Marwari breed of sheep over the other breeds studied in maintaining productivity under desert conditions. Prolonged intermittent (twice weekly) watering had no adverse effect on animal production in terms of body weight, wool growth and lambing performance of ewes of the Marwari and Magra breeds. Water restricted (watered twice weekly) animals of both the breeds consumed less than half the quantity of water consumed by the daily watered groups. Thus, with a flock of 100 adult sheep, about 6,500 liter of drinking water could be saved per month by resorting to a twice in a week watering schedule. The daily watering of sheep would, therefore, appear to be a dispensable practice at least in the management of desert adapted breeds.

In one study one group each of Marwari goats and Marwari sheep was watered every 5th day and the other group was watered *ad lib* daily. The sheep and goats of group one lost 7% and 4% weight, whereas, body loss was 10% and 18% in respective breed between November and March, The sheep and goats of group one consumed water at the rate of 3.4% and % of their body weight per day, while the sheep and goats of second group consumed water at the rate of 8.5% and 9.2% of their body weight per day, respectively. Blood haemoglobin, PCV (%), total plasma proteins, albumin, globulin, A/G ratio, inorganic phosphorus, and urea; and urine electrolytes like Na, K have remained near normal in sheep and goats of first group. The wool and hair production also in this group did not change. No difference was observed in wool and hair yield in the two groups showing the utility (conservation of water) of watering every fifth day.

• *Responses of shorn and unshorn sheep to heat exposure*

Respiration rate of shorn animals was generally higher in the morning and afternoon. During hot period reverse trend was observed, as it was higher in unshorn animals. Rectal temperature was 38.9°C in the morning and increased to 40.2°C during noon followed by slight decline to 39.8°C in the afternoon. The skin temperature of the shorn animals was generally higher than their unshorn counterparts and the differences were marked during noon. During peak hot period although the respiration rate of unshorn animals was higher than the shorn counterparts, but their rectal temperature was lower indicating better heat tolerance (Kaushish *et al.* 1995).

• *Responses of sheep to solar exposure*

In one experiment hoggets of native Nali, Chokla and their crosses with Rambouillet and Soviet Merino were used to study the effect of solar exposure. Three animals from each breed were maintained both in shade and sun during summer. After first recording, the animals of sun and shade groups were exchanged and the observations were continued. The respiration rate, rectal temperature, skin temperature and fleece temperature at 1 and 2 cm above the skin were higher in the solar exposed group as compared to the animals maintained in shed and higher at 16 and 13 hr than at 9 hr. In general the rectal temperature of the crossbred animals was lower than the native (lower in Merino x Nali); and highest in Nali. It was concluded that the animals maintained under hot sun experienced greater stress as compared to the animals under shade. It also appeared that in crossbreds the thermolytic mechanism was activated at a lower ambient temperature resulting in their higher respiration rate and lower rectal temperature (Kaushish *et al.* 1998).

3.4 Goats

The important goat breeds of arid Rajasthan are Marwari, Parbatsar, Jhakrana, Sirohi and Shekhawati. Mehsana and Kachchhi breeds are found in the arid region of Gujarat. The goats are primarily raised for meat purpose. The Parbatsar and Jhakrana breeds are known for their reasonably good milk and meat production. Marwari doe on an average yields 84 litre of milk in 180 days lactation period, whereas Parbatsar doe may yield up to 132 litre in 170 days lactation period (Patel *et al.* 1999). The goat is extremely well adapted to the arid environment. It grows faster, breeds more efficiently, can tolerate higher salt loads, needs less water and has liking for a wider variety of feeds including many weeds, as compared to the sheep. The fact that the goat is mainly a browser and the sheep a grazer indicates a relevant role of the former in the desert ecosystem. The goat is not necessarily the most important biotic factor involved in desertification. The deep-rooted bias against the goat may not be wholly justified on scientific grounds and its proper place in the agro-system of the desert needs to be redefined, particularly in view of the goat's potentialities to meet the protein gap in the country. The studies on productivity and adaptability of various goat breeds (e.g. Marwari, Parbatsar, Shekhawati, Jhakrana, Kutchhi, Barbari and Jamunapari) under desert conditions showed that these breeds were non seasonal in their reproductive behavior. The incidences of oestrus were very low in case of Jhakrana and Jamunapari goats (Mathur and Mittal 1990).

• *Effect of supplementation on productivity of desert goats*

Under existing small ruminant production system the slaughter weight of lambs and kids in the country is lower and age at which it is usually achieved is much higher. The system of raising finisher lambs and kids for meat purpose under grazing with diet supplementation, is cost effective for the producer and has not been largely adopted by the farmers owing to their poor economic background and restricting to their age-old traditional practices. The experiment conducted in the country on performance of sheep or goats under grazing with supplementation has been periodically reviewed (Patanayak and Singh 1972, Patnayak and Mohan 1974, Patnayak 1993, Karim 1995 and Karim *et al.* 1996). The experiment results indicated that in addition to free grazing on rangeland, limited amount of concentrate supplementaion (1.5 to 2.0% of body weight) would provide 25 and 30 kg finishing weight at 6 and 9 months of age, respectively. Grazing with supplementation on established pasture has potential of still higher production traits. The established *Cenchrus* pasture stocked as per carrying capacity with concentrate supplementation 1.5% of body weight provided 27.3 and 26.2 kg finishing weight in lambs and kids at 6 months of age, respectively (Shinde *et al.* 1995). However, grazing with *ad libitum* concentrate supplementation had no added advantage in the kids as they achieved 25 kg at 6 months and 30 kg at 9 months of age even under regulated level concentrate supplementation.

4. Some Technologies / Tools for Improving Livestock Production During Drought

- **Drinking water sachedules for livestock:** Water is a scarce commodity in arid region; the problem has become more acute during last 25 years. Studies conducted revealed that desert sheep can be kept on twice weekly watering schedule for their entire life without any apparent ill effect on their production. Meat goats in this region can also be kept on this water schedule. However, the thrice a week watering is optimum for satisfactory milk production in milch goats. By adopting this schedule at least 50% of water can be saved. Salinity in underground waters is the common occurrence in desert region. It is clear from 10 years in depth studies that desert sheep and goats can be reared on moderately saline waters containing 5 g l^{-1} of dissolved salts for prolonged period of time without any adverse effect on any production parameters.

- **Improving low quality fodder with urea treatment:** As such wheat straw is a very poor quality fodder because of its low palatability, digestibility, protein (CP 0.5%) and mineral contents along with poor intake. Efforts are being made for improving its nutritive values. Fodder enrichment with urea at 4% level increases CP value by 8% and simultaneously significantly increases its palatability and digestibility.

- **Economical feeding of cattle utilizing cheaper non conventional - *tumba* seed cake:** The cattle feed having 25% *tumba (Citrullus colocynthis)* seed cake has no adverse effect on the productive and reproductive performance of Tharparkar cattle (Mathur 2004). It is locally available in abundance (having 16-20% CP) and is a cheaper source of protein and is normally used as a fuel for furnaces and thus wasted. Monetarily it may save around 18-20% of cost without impairing the productivity. *Tumba*-seed cake is produced in large quantities annually. The oil is also extracted from it at most of the district headquarters of the state.

REFERENCES

Acharya R.M. 1982. Sheep and goat breeds of India. FAO Rome pp.19-50.

Burke, S. 1998. *Windbreaks*. Butterworth-Heinemann, Woburn, Massachusetts, USA.

Karim, SA. 1995. Nutritional aspects of meat production. Compendium 1 (Review papers). VII th Animal Nutrition Research Workers Conference, Bombay, pp.86-94.

Karim, SA, Shinde AK and Sankhyan SK. 1996. Prospects of sheep and goat for meat production in eastern region of the country. In: Workshop *on Prospect and Strategies for Sheep and Goat Production and Utilization for Eastern Region*. Bhubaneswar, 9-11 October.

Kaushish, SK. 1991. Improving sheep production through improved managerial practices. *Carpet e-world International* 13: 47-50

Kaushish, SK 1992. Carpet wool production and utilization in India. *Carpet e- world International* 14: 70-74.

Kaushish SK, Bohra HC, Mathur BK and Patel AK. 1998. In: *Fifty Years of Arid Zone Research in India*. (eds.). Faroda AS and Manjeet Singh Pp.393-426.

Kaushish SK, Karim SA and Rawat PS. 1995. Physiological response of unshorn and shorn native and crossbred sheep under heat exposure. *Indian J. Animal Sci*. 65: 714-717.

Kaushish S.K. 2006. Sheep breeds of India and Their production. In: Stratigies for Sustainable Livestock Production (ed. Kaushish, S.K., Bohra, H.C., B.K Mathur Patel, A.K and A.C. Mathur. Kalyani Publishers Ludhiana. pp 323

Kaushish SK, Rawat PS and Sharma SC. 1990. Performance of native sheep (Malpura) and its crosses with Avikalin under semiarid condition. *World Review of Animal Production* 25: 44-46.

Mathur BK and Mittal JP. 1990. Productive performance of various goat breeds on natural protected rangelands in arid region. In: National Seminar on *Physiological Status of Indigenous Livestock, Its Contribution to World Rural Eeconomy as Prospects,* held at Assam Agricultural Universities, Khanapara, Guwahati, 24-26th October. pp. 17.

Mathur, B. K. (2004). Drought proofing through utilization of low – grade roughage and non-conventional feed resources for small ruminants. National Seminar on" *Opportunities and Challenges in Nutrition and Feeding Management of Sheep, Goat and Rabbit for sustainable production*". (February, 10-12).organized by Indian Society of Sheep and Goat Production and Central Sheep and Wool Research Institute Avikanagar-304501, Rajasthan. p. 191-201.

Patel AK Mathur BK and Kaushish SK. 1999. Comperative productive performance of goats in arid rangelands. *Indian Journal of Animal Sciences,* 69:139-140.

Patel AK Mathur AC and Kaushish SK. 1997. Factors affecting production and performance of Tharparkar cattle under semi arid condition. In:

Proceedings *Recent Advances in Management of Arid Ecosystem* held at CAZRI, Jodhpur. pp. 145.

Patnayak BC. 1993. Nutrition of grazing ruminants In: *Recent Advances in Ruminant Nutrition*. CSWRI, Avikanagar.

Patnayak BC and Mohan M. 1974. Performance of crossbred lambs on supplementation of concentrate and cow pea hay. *Indian Journal of Animal Sciences* 44: 376.

Patnayak BC and Singh M. 1972. Sheep nutrition research in India. In: Review of research on sheep and wool production in India. CSWRI Monogram Publication, pp.49.

Shinde AK Karim SA Singh NP and Patnayak BC. 1995. Growth performance of weaner lambs and kids under intensive and semi intensive feeding management. *Indian Journal of Animal Sciences* 65: 630-633.

25

Impact of drought on reproduction and fertility in Animals

Alok Chand Mathur

Krishi Vigayan Kendra, Central Arid Zone Research Institute, Jodhpur - 342 003

Drought occurrence in arid region of Rajasthan is a regular phenomenon. It is used to be a typical tropical drought, being associated with high ambient temperature. In spite of recurrent droughts, animal husbandry is the backbone of arid rural economy being blessed with best drought hardy breeds of cattle, sheep, goat, horse and camel. Further, the livestock population is continuously increasing as a result; there is tremendous pressure of livestock on land.

In a broad term, applicable for a livestock keeper, drought is lack of sufficient precipitation at the time of year in which plants are dependent upon the moisture for growth. As a result, there is drastic decrease in forage production, both in terms of quality as well as quantity. The tropical drought is always associated with direct effect of heat-stress on animals apart from fodder scarcity conditions.

Plant response to drought

Leaf and shoot growth slows as soil water becomes less available. Tillering in grasses is also reduced. As drought condition persists, leaves wilt, fold, discolored and may eventually die. Thus there is drastic reduction in total amount of plant biomass availability. During drought plant utilize carbohydrates that were produced the previous growing season and stored in the roots or crown of the plant. This results in decrease in carbohydrate content of the already reduced biomass. This drain on stored carbohydrate result in loss of root vigour and fewer basal buds develop for the next year growth. Thus, drought can result not only in reduction in forage in the year of drought but also in subsequent years, even if rainfall is adequate.

Affect of drought on animal performance

Low forage production results in livestock having to range further to obtain necessary amounts of dry matter to meet their nutritional needs. These extra efforts result in the expenditure of additional energy that may result in a loss of body condition for mature animals and reduced growth rate in younger stock. Walking in search of feed and water can use up as much as 30% of the energy a cow derives from feed (McDougald et al., 2001).

If a cow or ewe is not able to obtain sufficient nutrients, particularly energy, she will loose considerable weight. A loss in body condition of a lactating cow or ewe results in a decrease in milk production and reproductive activity may cease. Body condition scores (BCS) of a cow 4-10 yrs of age needs to be above 4 and for a ewe 3-7 yrs of age and above 3 at time of parturition and breeding otherwise they will have lower milk production and fertility resulting in birth of underweight offspring and in fertility. For survival loss of body weight upto 20% in cattle and 30-40% in arid sheep and camel is reversible. In general, animals that loose 30% or more of their normal body weight will nearly always die (Young and Scrimshaw, 1971).

Effect of drought on Reproductive potential

In nature it is general rule that animals breed annually and for the first time when it is attain a specific body weight. Parturition occurs when food for the mother is abundant to ensure adequate lactation for the nourishment of the offspring. Under the condition of housing and feeding provided by domestication, the breeding season tends to be lengthened as a result animals (bovines) may breed at any time of the year. All domesticated animals however show a tendency to revert to the natural breeding season (Arthur et al., 1989). Recurrent droughts of the arid region, force the animals to revert to the natural breeding rules i.e. 'improper nutrition- no breeding' to a certain extent.

In farm animals successful reproduction is considered gateway to milk production from animals as lactation follows successful reproduction only. Hence optimization of reproduction is the key for profitability from animals. Normal fertility, in general sense as applicable to cattle, denotes one viable calf each year. Infertility is simply a degree of reduced fertility.

Infertility in large animals is the result of several factors which include genetics, nutrition, environment, herd health status and production potential of the cow herd. The task of minimizing the negative effects of these factors on fertility is compounded because they do not act independently of each other.

Drought affect the reproductive performance of animals by the interaction of the above factors only, mainly through prolonged conditions of negative nitrogen balance and of heat- stress.

The important causative agents of the adverse effects of drought on reproduction are as follows.

1. Malnutrition
2. Environmental stress
3. High incidence of parasitic infestations
4. Higher incidence of diseases

(I) Effect of Malnutrition on reproductive potential- as discussed above drought has direct effect in lowering the nutritional status of animal i.e. malnutrition. Nutrition is the most critical factor affecting reproduction. To be specific, drought-affected forage biomass contain lower concentration of Crude Protein, Energy, Vitamin- A and Phosphorous along with micro-elements. All these nutrients have direct role in maintaining normal reproductive potential.

Most of the field infertility cases in the arid region, especially during drought years, are usually due to multiple deficiencies of the above nutrients compounded with environmental heat- stress which directly ceases ovarian cyclicity.

The initiation of puberty is largely a function of the animals age and maturity since the female is born with a genetic potential for cyclic reproductive activity. Normally the females of domestic species reach the age of puberty at a specific age who in turn is closely related to body weight. Thus, the animals that are well fed with good growth rates reach puberty before those that are poorly fed with slow growth rates. However, unless the animal is severely malnourished the onset of cyclic activity will eventually occur (Arthur et al., 1989). Delayed skeltol maturity resulting from underfeeding can result in decreased pelvic width leading to dystokia which may further lead to increased post-partum interval and decreased conception and pregnancy rates (Walker et al., 1994)

(a) Protein Deficiency: Prolonged drought condition lead to hypoproteinemia which otherwise, is also a condition of common occurrence in cattle, reported as infertile cases in rural arid areas. It is commonly associated with inanition and starvation condition of cattle. The effect of low level of protein intake on reproduction may be due to reduced intake of total feed resulting in delay in oestrous (Wiltbank et al., 1965).

(b) Hypoglycemia: Reduced level of energy in animals during drought condition is common. It is due to existing lower levels of energy in the available forage biomass and additional energy loss by animals in search of such biomass in otherwise harsh arid environmental conditions. The level of energy is considered more important for reproduction than is protein. When the energy in the adult animal is low, follicle fail to develop to maturity resulting in follicular atresia along with a loss of sexual desire and anoestrous. (Roberts, 1971).

Hypoglycemia at estrus and shortly after service may exert a harmful effect on conception (Mcclure, 1968). This effect may be brought about by lowering of glucose or glycogen levels in the mucosa of genital tract resulting in a lack of energy for spermatozoa or fertilized ova.

(c) Vitamin-A deficiency: Clinical symptoms of Vitamin-A deficiency are wide spread in arid region in general, but are pronounced during drought. The animals usually suffer from acclimation, night-blindness, debility, thriftiness of skin coat along with abortion and still-birth incidences. There is controversy regarding role of Vitamin-A with anoesterous condition. The deficiency effect occurs during later half of gestation and is characterized by abortion or by the birth of weak or dead calf. Vitamin-A deficient animals may have normal oestrous cycle, ovulate and conceive with normal early fetal development even though epithelial or other tissue changes have developed (Roberts, 1971).

Vitamin-A deficiencies results in degenerative changes in epithelial tissue like keratinization and degeneration of the placenta which may lead to fetal death, abortion, dystokia, retained placenta and septic metritis.

Photo 1. Examination of gynaecological cases under rural conditions.

Though β-carotene is the plant precursor of Vitamin-A, recent studies have shown separate importance of β-carotene and Vitamin-A. Bovine luteal tissue has one of the highest concentrations of β carotene as compared to any other tissue but controversial role of β - Carotene deficiency in cases of delayed estrus, silent estrus, follicular cyst, LH and progesterone concentrations have been reported (Arthur et al., 1989).

But based on the response to the line of treatment of true anoestrous cases in drought affected rural arid region by us, Vitamin-A supplementation is considered very important in dealing with such infertility cases (Mathur et al., 2001). McDougald et al. (2001) have also considered Vitamin-A supplementation as critical during drought.

(d) Phosphorous deficiency: Phosphorous deficiency symptoms are quite common and widely spread in animals of the arid areas. The symptoms include inanition, poor hair-coat, depraved appetite and infertility in form of delayed puberty, silent estrus, post-partum anoestrous with ovarian acyclicity or sub-dued activity. Source of phosphorous deficiency is considered to be the arid soil.

Phosphorous supplementation through inorganic/organic salt injectables and mineral mixtures in such infertility cases in the region has saluting effect. Hence, Phosphorous supplementation has been an important constituent of package for treating true anoestrous in the bovines of rural arid region (Mathur et al., 2001).

Role of Phosphorous in successful normal reproduction is well documented (Morrow, 1969; Read et al., 1986ab) but the evidence for the importance of hypophosphatemia as a cause of infertility is conflicting (Arthur et al., 1989). Morris (1976) suggested that a blood Phosphorous level of less than 4mg/dl in infertile animals at its peak production level confirms the diagnosis of hypophosphatemia as cause of infertility. Meeting out the normal phosphorous requirement for body maintenance of 13 gm day (requirement for milk production extra) from the drought affected forage biomass of the already phosphorous deficient arid soil seems to be unexpected.

(e) Trace minerals deficiency: Specific Trace minerals namely Copper, Cobalt, Manganese, Iodine, Selenium and Chromium are considered to have role in maintaining reproductive potential but their direct role in the reproduction process is a subject of contradicting reports. Improvement in the fertility by their supplementation to infertile animals speaks of their importance. The deficiencies of these trace minerals are considered much higher in arid areas especially during drought conditions.

(II) Effect of environmental stress on reproduction during drought: In majority of arid areas animals are mostly maintained on extensive grazing on range lands where these animals are not provided any kind of shelter during grazing browsing except of natural trees. The average grazing period in arid region is about 10hrs/day. It has been estimated (Patel, 2003) that about 80% arid farmers provide thatched house for cattle and buffaloes while only 10% farmers provide thatched house for sheep and goats during evening and nights after return from grazing.

Thus under the normal managemental conditions the animals under arid region are subjected to high degree of heat-stress involving solar radiation, wind velocity, air temperature and humidity. Livestock being homeotherms looses lot of energy under such stress conditions to maintain homeostasis through direct involvement of

Central Nervous System and endocrine system which are directly involved in controlling reproduction process. Thus there is direct effect of heat- stress on hypothalamus- pituitary axis which in turn controls gonadal activities.

(a) Heat-stress and fertility: Heat stress is the major contributory factor to the lowered fertility in bovines affecting both male and female fertility. The conception rate in cattle are 20-30% lower in hot season (compared to winter season) along with lowered incidence of estrus detection, day to first service (Cavestancy et al., 1985; Badinga et al., 1985; Rensis et al., 2002; Almier *et al.,* 2002).

It has been suggested that there could be longer lasting effect of heat stress of summer months on the antral follicles that will develop into large dominant follicles 40-50 days later (Roth *et al.,* 2001 ab). In general, heat-stress reduces the duration and intensity of estrus in dairy cattle (Rensis and Scaramuzzi, 2003). Reproductive efficiency is well known to be decreased during summer months both in male and female buffaloes (Bhattacharya *et al.,* 1970; Pant and Roy, 1972; Luktuke *et al.,* 1973; Sengupta and Bhela, 1988).

(b) Effect of heat stress on reproductive hormones: There is direct effect of heat stress on hypothalamus pituitary axis controlling gonadal activities. The effect of heat-stress on LH concentrations in peripheral blood is inconsistent. It has been concluded (Rensis and Scaramuzzi, 2003) that in summer the dominant follicle develops in a low LH environment which results in lower estradiol secretion from the dominant follicle leading to poor expression of estrus and hence reduction in fertility.

FSH is increased by heat-stress, may be due to decreased plasma Inhibin production by the compromised follicles. It is indicative of reduced folliculogenesis since a significant proportion of plasma Inhibin comes from small and medium sized follicles (Wolfenson *et al.,* 1995; Palta *et al.,* 1997).

The effect of heat stress in reducing plasma estradiol is clear but its effect on plasma progesterone is controversial. The effect of heat stress in reducing secretary activity of follicles (hence reduction in estradiaol concentration) leading to summer infertility is perhaps through increased corticosteroid secretion which can inhibit GnRH and LH secretion (Gilad *et al.,* 1993).

(c) Effect of heat stress on Follicular development: Recent evidence indicate that the development of oocytes is also temperature sensitive (Ruteledge et al., 1999) indicating possibility of high ovarion temperature during heat stress. Heat stress delays follicle selection and lengthens the follicular wave and thus has potentially adverse effects on the quality of oocytes and follicular steroidogenesis (Roth et al., 2001ab). Summer heat stress reduces the degree of dominant follicle and more medium sized dominant follicles survive (Woolfenson et al., 1995) as a result more than one dominant follicle can develop leading to increase in twinning rate (Ryan and Boland, 1991)

(d) Effect of heat-stress on embryo survival: Heat stress has reported to result in decrease in blood flow to the uterus and increase in uterine temperature (Roman-Ponce et al., 1978) which inhibit embryonic development leading to early embryonic mortality but the magnitude of the effect decrease as embryo develops. Most embryo loss occurs before day 42 in heat stressed cows (Ealy et al., 1993; Vasconcelos et al., 1998). Heat stress can affect endometrial prostaglandin secretion leading to premature luteolysis and embryo loss.

Thus heat stress has a wide range of effect on the reproductive axis. Some of these effects directly affect individual reproductive organs such as hypothalamus, anterior pitutary, the uterus, the follicles with oocytes and the embryo itself. The indirect effect of heat stress is resulted by the reduced dry matter intake and prolonged negative nitrogen balance. Thus the overall effect on fertility is due to accumulation of the effects of the above several factors (Rensis and Scaramuzzi, 2003).

(III) Effect of drought on Parasitic infestation: Cattle of all ages under nutritional and heat stress are resistant to parasites than under normal condition. It is applicable for both internal and external parasites hence strategic deworming during a drought will relieve some of the nutritional stress on the animal (Hupp and Rathwell, 1998). There is already a higher load of parasites in the arid livestock due to availability of scarce and unhygienic water supply from the common rural ponds.

These high incidences of endo and ectoparasitism cause further precipitation of nutritional deficiencies in the already underfed animals during drought. Hence it is advisable that before start of any nutritional supplementation programme to combat infertility during drought, control of both endo and ectoparasites, should be taken up as a pre-requisite step in the arid region.

IV. Disease incidence during drought affecting reproduction : Any general specific infectious or noninfectious disease incidence occurring during drought lowers the nutritional status and general resistance of the animal which in turn affect the reproduction process to varying extent. Such affections are

(a) Infectious diseases : As cattle graze on shorter and shorter forage during drought, the chance of picking up soil- borne pathogens like Clostridial diseases (Black Quarter and Enterotoxemia), Leptospirosis and Anthrax increases (Ron Gill). Under prevailing Indian conditions incidences of HS and FMD may also be kept in mind.

(b) Non-infectious diseases (i) Deficiency diseases: Ketosis, Anaemia, Pica, milk fever, Grass tetany, Xerophthalmia, Ricketts, Osteomalacia, Goitre, Parakeratosis, Hypocuprosis etc.

(ii) Malnutrition linked diseases: Stomatitis, Gastroenteritis, indigestion, Tympany, Hepatitis, Nephritis and urolithiasis, Post-parturient Hemoglobinuria (hypophosphatemia), Choke, Traumatic Reticlo Peritonitis, dehydration etc (ICAR).

(iii) Heat Stroke/Sun Stroke: The condition may be sudden preceded by ataxia, weakness, hot skin, rapid pulse and high temperature (upto 107° F) and death (ICAR).

(iv) Poisoning and Toxicity: Starvation condition may lead to eating of poisonous toxic material like braken fern, lantana, camara, Rati (*Abrus prectorus*), Dhatura, Kner, cyanogenic plants like immature sorghum, maize and cereals affected with ergot, India Pea, Nitrate and Nitrite containing plants (I CAR).

Strategies to moderate impact of drought on livestock production and fertility

(I). Culling of animals: In conditions of threatened drought, attempt should be made to reduce the livestock numbers in a planned systematic method. Dry and unproductive animals need to be culled first. Destocking of animals shall conserve the limited feed and fodder resources for the more productive animals. Culling policy during drought has been mentioned (Blaine) but its practicability is difficult under Indian conditions due to religious and cultural compulsions.

(II) Segregation of animals: As heifers and calves are not able to compete with mature cows for pasture or supplemental feed, their segregation into suitable groups shall provide them a better chance of getting needed feed supplies. Vulnerable classes can be segregated and given preferential treatment. The older cows can be moved to the poorer forage fields. (Hupp and Rathwell, 1998).

(III) General Livestock Managemental Practices: The important managemental practices which are to be adopted during drought condition are given below (ICAR)

- Animal should not be kept in direct sunlight; they should be tied under a tree or in a shed.
- They should be allowed to graze only in early morning or late evening. Supplementation of feed and fodder should also be done during this period to decrease water requirement.
- There should be no overcrowding of animal in shed and animals should be provided proper ventilation.
- Roof of the shade should be better made of asbestos sheet. Roof made of iron sheets need to be covered with thatching material.
- Upper layer of Kaccha floor should be replaced with new sand in every six months so that load of organism can be reduced.
- Pakka floor should be properly cleaned and as per need disinfectant may also be used.
- Wild animals like fox, dogs, jackals who are responsible for spread of some infections like rabies, anthrax. FMD, HS etc. need to be controlled.

IV. Feeding strategies during drought- During drought malnutrition lead to loss of body weight. In case of milch and other producing animals, the only solution lies with supplementation because productivity loss cannot be recouped by any other mean.

The strategy in this regard may include -

- Storage in fodder banks
- Urea treatment of straws
- Supplementation of Vitamin-A and mineral mixture . Complete feed formalities
- Use of dry and fallen tree leaves.
- Use of unconventional feed resources.
- Water management.

V. Preventive measures against epidemics and diseases

(I) Control of endoparasites - by deworning with broad spectrum anthelmentics like Albendazole, fenbendazole or Oxefendazole atleast twice a year.

(II) Control. of ectoparasites - like ticks, mites etc. by regular spray of Deltamethrin (0.2% solution) or Cypermethrin (0.1% solution). For spray in the shed, the concentration needs to be at least 5 times more. The usual precautions for use of insecticides must be taken.

(III) Vaccination against prevalent infectious diseases like FMD (Polyvalent FMD Vaccine), HS (HS oil Adjuvent Vaccine) Anthrax (Anthrax spore vaccine), BQ (BQ Polyvalent Vaccine) Enterotoxemia (ET Vaccine), Sheep/goat Pox, PPR in goat (IVRI Vaccine)etc.

(IV) Coccidiostat- All calves at 0-3 months should be given Coccidio stat, malnly sulfonamides @ 220mg/kg b.w. on first day and @ 110mglkg b.w. for the next four days should be given (ICAR).

(VI) Management of heat stroke

(a) The affected animal should be moved to shade and Ventilated areas.

(b) Water should be poured on the body.

(c) Fluid therapy to check dehydration should be done in consultation with Vet. However, self- made preparation containing 10 litres of cold water properly mixed with 0.5 litre molasses or 0.5 kg jaggery, 20 gm common salt 0.5 kg oat or barley flour can be given orally twice daily for 2-3 days (ICAR).

(VII) Poisoning and Toxicity control

Locate all areas with poisonous plants and monitor them closely and prevent their access to the grazing animals. In general, toxic symptoms are seen in many animals of the flock/herd with specific symptoms which require specific antidote, several fluid therapy and injectable atropine sulfate through veterinarian.

REFERENCES

Almier, M.; Rosa, DG.; Grasso, R.; Napolitania, F.; Borde, A. (2002) Effect of Climate on the response of three *oestrous* synchronization techniques on lactating cows. Anim. Reproduction Sci 71:157-68

Arthur, GH; Noakes, DE and Pearson, H (1989) Veterinary Reproduction and Obstetrics VI ed. ELBS.

Badinga, L; Collier, RJ; Thatcher, WW; Wilcox, CJ (1985). Effect of Climatic and management factors on conception rate of dairy cattle in subtropical environment. J.Dairy Sci., 68:78-85

Bhattacharya, AR; Luktuke, SN and Roy, SN (1970) Incidence of normal and pathological conditions of She buffalo genetelia in different months. Indian J. Anim. Sci. 40:425

Blaine, E Animal performance under drought: what should culled File/ E/ Reproduction performance.htm

Byers, JH; Jones, JR and Bone, JF(1956) J. Dairy Sci 34, 1200.

Cavestany, D; EI-Whisky, AB; Foot, RH (1985) Effect of season and high environmental temperature on fertility of Holstein Cattle. J. Dairy Sci 68:1471-8.

Ealy, AD; Drost, M; Robinson, OW; Britt, JH (1993) Developmental Changes in embryonic resistance to adverse effects of maternal heat stress in cows. J. Dalry Sci 76: 2899-905.

Gilad, E; Meidan, R; Berman, A;Graber, X; Wolfenson, D (1993). Effect of heat stress on tonic and GnRH induced gonadotrophinsecrction in relation to concentration of estradiol in plasma of cyclic cows. J. Reprod, Festil. 19:315-21.

Gimenez, Jr. Maintaining Herd Productivity during drought

Horn, BE Animal Performance under drought: what should culled performance. htm.

Hupp, HD and Rathwell (1998) Herd Management during drought. Mangt. Marketing Memo. No. 369. htm p2.

Jolly, PD; Mc dougall, S, Fitzpatrick, LA, Maemillan, KL, Entwhisle, K (1985) Physiological effects of under nutrition on postmartum anoestrous in cons. J. Reprod Festil. Suppl. 49:477-92.

Luktuke, SN, Bhattacharya, AR, Singh, SK and Khan, BU (1973). Studies on aberration in functional activity of ovaries in buffaloes. Indian Vet. J. 50: 875-881.

Mathur, AC, Mathur BK, Kaushish, SK and Patil, AK (2001). Effect of package of improved nutritional management practices on true anoestraus cases in buffaloes in rural arid Rajasthan. Abstracts Souvenir XIX Ann. Conv. & Ntl. Symp. `Fertility management of Farm animals under adverse agro-climate condition' 6-8 October, 2001, Jodhpur (India): P. 39.

McClure, T.J. (1968). Hypoglycemia an apparent cause of infertility of lactating cows. *Brit. Vet. Jour.* 124: 126.

McDougald, NK; Frost, WE and Clawson, WJ (2001) Livestock management during drought, Rangeland Management Series, University of California Publication No. 8034.

Morrow, DA (1969) Phosphorous deficiency and infertility in dairy heifers. JA VMA 154:7:761

Palta, P; Mandal, S; Prakash, BS; Madan, ML (1997) Peripheral Inhibin levels in relation to climate variations and stage of *estrous* cycle in buffalo (*Bubalus bubalis*). Theriogenology 47:898-995.

Pant, HC and Roy, A (1972). The water buffalo and its future. In: Improvement of livestock productivity in warm climate (ed. Me Dowell RH) WH Freeman and Co, San Fransisco pp 563-599.

Patel, A.K. (2003). Role of shelter management in improving production under arid condition. Winter school on `Strategies for Sustainable Livestock Production in arid Zone' 29 October - 17th November, 2003, CAZRI (ICAR), Jodhpur (India) P. 22-26.

Read, M.P; Engels, EAN and Smith, W A (1986a) Phosphorous and the grazing ruminant. The effect of supplementary phosphorous on sheep at Armoedsvlakte South African J. Anim. Sci 16: 1-6.

Read, MP; Engels, EAN and Smith, W A (1986b) Phosphorous and the grazing ruminants 1. The effect of supplementary phosphorous on cattle at Glen and Armoedsvlakte. South African J. Ani. Sci. 16:7- 12.

Rensis, FD and Scaramuzzi, RJ (2003) Heat stress and seasonal effects on reproduction in the dairy cow- a review Theriogenology 60: 1139-1151.

Rensis, FD; Marconi, P; Capelli, T; Gatti, F; Facciolango, F; Franzini, Setal (2002). Fertility in postpartum dairy cows in winter or summer following estrous synchronization and fixed time AL after the induction of an surge with GnRH or hCG. Theriogenology 58:1675- 87.

Roberts, S.J. (1971) Veterinary Obstetrics and genital diseases II ed. CBS Publishers, New Delhi.

Roman-Ponce, H; Thatcher, WW; Canton, D; Barron, DH; Wilcox, CJC (1978) Thermal stress effects on uterine blood flow in dairy cows. J. Ani. Sci. 46: 175-80.

Ron Gill Maintaining Herd performance during drought. The Texas A&M Uni. System AS web-018/8-99.

Roth, Z; Arav, A; Zeron, Y; Braw-Tal, R; Wolfenson, D (2001b) Improvement of quality of oocytes collected in the autumn by enhanced removal of impalred follicles from preovulatory heat stressed cows. Reproduction 122:737-44.

Roth, Z; Meweidan, R; Shaham-Albalancy, A; Braw-Tal, R; Wolfens, D (2001a) Delayed effect of heat stress on steroid production in medium size and pre ovulatory follicles. Reproduction 121:745-51.

Ryan, DP and Boland, MP (1991) Frequency of twin births among Holstein Friesian cows in a warm dry climate. Theriogenology 36: 1-10.

Sengupta, BP and Bhela, SL (1988) current status of male fertility research in buffaloes. II world Buffaloes Congress Vol. II pp 229- 243.

Vasconcelos, JLM; Silcox, RW; Lacerda, JA; Pursley, GA, Wiltbank MC (1998). Pregnancy rate, pregnancy loss and response to heat stress after AL at 2 different time, from ovulation in dairy cows. Bios Reprod. 56 (Suppll):140.

Walker, D; Retchie, H and Hawkins, D (1994) Getting the cow herd bred MSU Extension Beef Bulletins *23290001* (E:htm).

Wiltbank, IN; Bond, J; Warwick, EJ; Davis, RE, Cook, AC; Reynolds, WL and Hazen, WH (1965). Influence of protein intake on reproductive performance in the beef female through second calving Tech Bull. 1314 Ag Res Service USDA Washington, DC.

Wolfenson, D; Thatcher, WW; Baldings, L; Mediden, BJ et al (1995). The effect of heat stress on follicular development during the oestrous cycle in dairy cattle Biol. Reprod. 52: 1106-13.

Young, VR and Scrimishaw, WS (1971) The Physiology of starvation. Sci Am 225:14-21.

26

Livestock management during drought

A.K. Gahlot

College of Veterinary and Animal Science, Rajasthan Agricultural University, Bikaner - 334 001

With an area of 3.2 lac km^2, the hot arid zone constitutes about 12% of the total geographical area of the country. Out of this, major portion (more than 60%) of area exists in Rajasthan. The arid regions are characterized by wide variations in rainfall, soil type, irrigation facilities and micro-ecological conditions. The frequent drought leads to complete failure of crops very often, so much so that a good crop year has been observed to have a cyclic nature of 11 years. It is, therefore, not inappropriate to conclude that the major stay of rural economy in hot arid region is still animal husbandry. Livestock rearing is the major component of arid agriculture. In addition to providing quality nutrition in terms of meat and industrial raw material (wool, leather etc.), the contribution of small ruminants to arid crop farming through manure cannot be underestimated. Small ruminant-based animal husbandry also enhances the economic viability and sustainability of farming systems, particularly in rain fed areas. Arid region of the country is certainly better placed than the rest of the regions in terms of meat (>20% in terms of live animals) and fiber (>25%). That may be the reason that the people in this area suffer less from energy-protein deficiencies and hunger despite the lowest crop production. Animal husbandry in arid region has the added advantage of stability, round the year employment opportunity, equity distributive and being technologically simple (Chaudhry *et al.,* 1999).

Drought is considered by many to be the most complex but least understood of all natural hazards, affecting more people than any other hazard. Lessons from developed and developing countries demonstrate that drought results in significant impacts, regardless of level of development, although the characters of these impacts differ profoundly. Thought, drought is a normal feature of climate and its recurrence is inevitable, yet, there remains much confusion within the scientific and policy community about its characteristics. It is precisely this confusion that explains, to some extent, the lack of progress in drought management in most parts of the world.

Arid and semi-arid lands have probably never been under so much pressure, from climate variability, climate change and the demands of ever-increasing livestock and human population. Climate variability is a feature of many arid and semi-arid lands, though only the most protracted periods without rain can correctly be described as droughts. Protracted droughts lead to pressure being put on vegetation and soils through over-stocking, and on governments to provide food or financial assistance. People living in the areas prone to drought can become more self-reliant through being proactive rather than reactive in coping with drought. However, an essential pre-requisite to achieving increased self-reliance and to reduce the impact of drought is to put a state drought policy in place and provide the necessary research and education infrastructure so that farmers, agribusiness and rural communities can better anticipate and cope with droughts.

As income and employment in the crop sector started diminishing, the non-crop sector, which included dairy and poultry, underwent a significant upward rise. With the demand for poultry increasing and production reaching 40 billion eggs and 1.2 billion broilers, this sector now employs around 1.6 million people. At least 80% of employment in the poultry sector is generated directly by these farmers, while 20% is engaged in feed, pharmaceuticals, equipment and other services required by the poultry sector. Additionally, there is a similar number of people who are engaged in marketing and other allied activities in the poultry sector. All the above illustrate the point that we have a gigantic potential in all of these sectors, which are potent employment and income generators for the rural masses. With skewed income and agricultural holdings being a reality of the country's rural land holding structure, livestock, fisheries and dairy development quite clearly hold the key for providing employment and income to large segments of the underprivileged. Increasing the purchasing power of the rural and periurban masses through the aforesaid activities would ensure that they could considerably improve their consumption and nutrition basket.

In 1950-51, the contribution of agriculture sector to the national GDP was 50.5% with the share of animal husbandry as 15.5%. The share of agriculture went down to 20.5% by 2003-04, and the animal husbandry still contributing 1/3rd of it. The current contribution of animal husbandry is estimated around Rs. 200,000 crores annually. The market value of Indian livestock is estimated at Rs. 65,000 crores exceeding the value of food grains produced, annually. Salient features of Animal husbandry sector of India are:

- India has about 16% of cattle, 57% of buffalo, 17% of goats and 5% of sheep population of the world.
- India ranks first in respect of cattle and buffalo, second in goats, third in sheep and sixth in poultry population.
- The livestock sector employs 30 million in principal and subsidiary sectors of which women make up 71% of the labour force.
- The contribution of livestock and fisheries sector to the total GDP during 2003-04 was more than 7%.
- The total export earnings from livestock, fishery and related products was more than 10,000 crores in 2003-04.
- The landless, marginal and small farmers have more than 80% of the bovine population with operable area of only 57%.

Livestock rearing is the major component of arid agriculture, with particular reference to the state of Rajasthan. In addition to providing quality nutrition as milk, meat, eggs etc. and industrial raw material as animal fiber, leather etc., the contribution of livestock to arid crop farming through drought power and manure also cannot be underestimated. Animal husbandry also enhances the economic viability and sustainability of farming systems. It diversifies production and management options; increase total farm production and income. Stable and consistent growth rate in animal husbandry sector will help in increasing employment and bring about income redistribution in favour of small and marginal farmers and landless labourers. It is required therefore, to make Rajasthan a leading player in livestock product markets through sustained and all-round improvements in quality and efficiency by enabling the small producer to gainfully participate in the process of growth and modernization of the livestock sector and simultaneously ensuring the ecological sustainability of livestock sector during growth and modernization. The sector has proved as a means to provide -

- stability with a consistent growth rate of over 6% per annum,
- round the year employment opportunity with more than 50% of total house hold income in arid region as against 22.5% of national average,
- equity distributive as indicated by the fact that the share of the bottom 60% house holds in the ownership of bovine has increased from 58.8% in 1982 to 65.7% in 1992.
- Technologically simple vocation as traditionally, the arid farmers keep livestock in proportion to the 'free' crop residues and family labour available in their own household production systems and convert these into food, fuel and farm power-making each house hold a virtual self contained production system with no purchased inputs and little marketed outputs. All this makes the arid livestock farming a technologically simple enterprise to be followed as a tradition.

Strengths

- Most important economic activity in arid zone (more than agriculture)
- In other agro-climatic zones it is next to agriculture
- More than 75 rural house holds have livestock as an important component
- Average income from livestock is 30-50% whereas it reaches 60-80% from dairying
- 90% of all labour is from women
- Women have no say in its appropriation
- Ownership highly dispersed in ESWS
- Minimum credit-NABARD finances only less than 25% to this sector
- 80% rural house holds
- Ranks 3rd in milk production in the country next to Punjab, which stands second.

- Produces 9% milk, 40% wool and 10% meat of the country
- Asset worth Rs. 25,000 crores
- Drought proofing technology
- Traditional
- Women empowering
- Labour intensive
- Rural self employment
- Technologically simple & adaptive
- Production worth Rs. 20,000 crores

Livestock Asset of the State (Yr. 2003)

SPECIES	IN LACS	TOTAL VALUE IN LAC Rs.	SURPLUS VALUE IN LAC. Rs.
CATTLE	109	1000000	03,500,00
BUFFALO	105	1100000	04,000,00
SHEEP	101	0175000	00,600,00
GOAT	168	0200000	00,700,00
CAMEL	005	COMPARATIVELY LITTLE	
OTHER	005	COMPARATIVELY LITTLE	
TOTAL	492	#2500000	#0880000

Worth of Milk Production 2002-03

SPECIES	IN '000 TONS	VALUE IN LAC Rs
COW	2450	2000,00
BUFFALO	4800	4800,00
GOAT	0850	0650,00
TOTAL	8100	7450,00

Worth of Poultry Production 2002-03

NUMBER	IN LACS	VALUE IN LAC Rs.
POULTRY	0075	004,50
EGGS	6000	060,00
BROILER	0900	045,00
TOTAL		109,50

THREATS TO ANIMALS IN ARID ZONE

1. Climate: Hot and dry climate used to be considered as a safe climate from the point of view of infectious diseases. Nevertheless, recently some of the diseases of humid climate like those of PPR, tuberculosis, contagious mastitis, schistosomiosis, high concentration of internal parasitism etc. have been recorded in the arid region also. Heat stress/exhaustion/stroke has become quite common. Conversion of hot dry climate into hot humid climate after incorporation of a vast area of western Rajasthan under canal/well irrigation is also a predisposing factor to many hitherto unreported stress (Singh *et al.,* 1994), infectious (Mittal, 1987) and parasitic diseases (Pathak, 1997).

2. Distant pastures: With degradation of earstwhile pastures (Mertia, 1987), half hearted attempts of their conservation and drastic reduction of Common Property Resources (CPR), the pastures have gone too away from the habitations. Animals still on complete/partial range rearing practice are more prone to nutritional deficiencies including protein energy as well as mineral vitamin (Fakhruddin, 1987). Furthermore, prolonged timings when the animals are at pasture predisposes them to attack of several insect vectors/mechanical transmitters of diseases (Gahlot, 1997).

3. Prevalence of important diseases: Arid zone is registering continuous and increased prevalence of many epidemiologically important diseases of animals, which is a serious threat to animal husbandry in this area. Diseases like FMD, rabies, contagious ecthyma, sheep and goat pox, tuberculosis, Johne's disease, enterotoxemia, mycoplasmal infections, coccidiosis, brucellosis, anthrax, rickettsial and chlamydial diseases, nematode infestations, mange, pica (Kishore, 1998), deficiency of Vitamin A and D (Fakhruddin, 1987), Polio-enaphalomalacia (Tanwar *et al.,* 1983; 1994), deficiencies of calcium, phosphorus, zinc, iron etc. are not only prevailing unabated but also are registering an upward trend. High incidence rate of sub clinical mastitis has also been reported from this area (Singh *et al.,* 1999).

4. Scarcities of feed and fodder resources: There is an evident scarcity of feed and fodder resources in arid zone. Some of the factors responsible for this deficiency, such as degradation of pastures, drastic reduction of CPRs and persistent drought resulting often into complete crop failure and consequently complete absence of crop residues. Most of the feed and fodder ingredients are now being imported from other areas and states. Some of the dietary ingredients have replaced the erstwhile *Lasiurus sindicus* straw and *Zizyphus numularia* leaves almost to the extent of 80%. Similarly, the traditional ingredients of concentrate ration, such as, sesame cake and Guar *(Cymopsis tetragonoloba)* seeds have almost been completely replaced by cotton seed/ground nut/mustard cake, legume churi (coarse ground industrial by product), DORB, wheat bran etc. This switch over has resulted into higher incidence of metabolic (production) diseases. Production diseases were earlier considered to be related to only high producing exotic/cross bred animals, particularly cattle and buffaloes. Their incidences in the indigenous small ruminant animals have proved that dietary factors have become important in this group also.

5. Deficiencies: Several deficiencies have been identified in the animals of arid zone (Dongre, 2000) because of evidence of existence of deficiency, deficiency as the cause of the disease and cure or prevention by correction of the deficiency. Energy and protein deficiency results in to pica. Copper, cobalt selenium, zinc (Singh *et al.*, 1994), iron and iodine are some of the trace minerals resulting into general conditions like anaemia, retarded growth, reproductive disorders, pica (Kishore, 1998) and certain specific conditions suggestive of deficiency. Apart from these, deficiencies of vitamins A, D3, E, B_1 (Tanwar *et al.,* 1983, 1994) and C have also been identified.

6. Toxicity: Cyanide, sodium chloride, nitrate and nitrite poisoning have been reported to occur in this zone since long (Sharma and Gahlot, 1997). Newer toxicity includes those with pesticides/ farm chemicals, feed additives, drugs, environmental pollutants etc. (Radostits *et al.,* 1994).

Pesticides / farm chemicals: With frequent use of pesticides/farm chemicals in agriculture in this zone, particularly in the areas under irrigation, risk of poisoning of livestock has increased many folds. This is mainly because most of the domesticated animals in such areas are reared as a component of mixed farming system. Accidental access to insecticides/pesticides/farm chemical is the main cause of poisoning in not only animals but also human beings. Chlorinated hydrocarbons like chlordane, toxaphene. heptachlor etc., organophosphorus compounds and carbamates like endosulphan, monocrotophos, chlorpyriphos etc. and chemicals like urea are among the commonest causes of poisonings. Seed dressings, rodenticides and herbicides are another class of miscellaneous chemicals to which animals are exposed accidentally. Pesticides like melathion, asuntol, pyrethrins, amitraz etc. are also used in control of ectoparasites of animals. Application of these compounds in the concentrations and applications higher than the recommended ones often results into toxicity (Radostits *et al.*, 1994).

Feed additives: Feed additives like monensin, tiamulin, salinomycin etc. are not in use currently in this area. The only feed additive, particularly to improve the quality of roughage, is urea. Accidental overdosing of urea occurs as a result of miscalculation of the concentration. Any amount of urea over 1-2% may cause urea toxicity in ruminants.

Drugs: With the availability of several newer drugs and compounds, narrow therapeutic index, wide spread contraindications, anaphylactic reactions and adverse drug reactions have increased the risk of drug toxicity in animals. Ivermectins, antibiotics, antiprotozoan drugs like dimenazine, quinapyramine etc.

Environmental pollutants: The major environmental pollutant affecting the health of farm animals, particularly cattle, is plastic and polythene. *Per se,* there is no toxicity of polythene, but owing to its nonbiodegradable property, it acts as a non penetrating foreign body in the rumen of ruminants, leading to complete cessation of fermentatory activity and motility of rumen. Such animals necessitate surgical intervention in the form of rumenotomy. Recovery in such animals is also comparatively slower due to long standing malnutrition and debility. Other environmental pollutants include industrial waste, lead, arsenic etc. Due to influx of feed and fodder ingredients in the arid zone from other areas, phytotoxins and mycotoxins have been frequently identified in feed and fodder. These toxins not only affect health directly, but also production and reproduction of farm animals. Prolonged and poor storage conditions have been identified as the most important contributory factors to mycotoxicosis. Apart from direct effect on the health of the animals, these have been found to cause hyperthermia and precipitation of the production diseases.

Strategies-Technology options

a) *Germplasm*: The state is endowed with vast livestock populations having wide genetic diversity suitable for milk, meat, fiber and draft power. While giving due emphasis for their genetic improvement for efficient production, there is need for conservation of this valuable germplasm. Critical evaluation and classification of various breeds may be carried out concurrently but germplasm conservation work need not wait for detained classification.

b) ***Feed resource***. In recent decades, growing concern has been expressed about the large and expanding gap between feed and fodder resource availability and demand to be an alarming 50-80% of the requirement. With such a glaring shortage of feed and fodder, how can an increasing livestock population allow increase productivity and total production at respectable rates? The reasons may be: farmers do not feed at scientific feeding rates; resources actually available as grazing, CPRs, forests and crop residues are greater than estimated. Nevertheless, it would be wrong to underrate the seriousness of the feed and fodder resource problem. It is difficult to find immediate solutions and therefore, research and development wings need to address this imminent problem urgently.

c) ***Livestock Management-Ecological aspect*** - For the livestock policy, it makes best sense to help generate alternatives that reduce the pressure on the rural people to pursue livelihood strategies detrimental to their habitat and natural resources; and promote symbiotic and synergistic relationships between local communities and their environment through participatory natural resource management.

d) ***Animal Health Management:*** Another critical constraint is the wide spread presence of a variety of animal diseases and the state's failure to eradicate/control them. This may not only affect livestock/product export, but also results in massive economic losses. Establishment of efficient animal disease surveillance, monitoring and forecasting on the state and regional levels and development of effective and economic vaccines and biological products using modern technologies to check the onset and spread of animal disease and pests will go a long way in supporting the growth of this sector.

e) ***Breeding Management*** through restructured A.I. programme comprising of input generation and delivery system.

f) ***Animal Production system:*** The state of Rajasthan has well-established cattle, buffalo, sheep, goat, camel and poultry productions. The research and development support need to be designed in a way that there is blending of the features of advanced input intensive technologies with traditional production systems. Backyard production systems/small holder livestock production farming particularly indigenous cattle, camel, donkey, poultry and goat needs special attention. Pig farming needs reintroduction into the state,

g) *Product processing*. HACCPS endorsed by ICMSF are required to be followed particularly in the meat processing industry.

REFERENCES

Chaudhry, AL.; Gahlot, AK. and Beniwal, B.K. (1999): Livestock resource management in arid region for food security and export promotion. Paper presented in the 4th Agriculture Science Congress, 21-14 February, 1999, B.M. Birla Science & Technology Centre, Jaipur (Raj).

Dongre, R.A (2000): Haemato-biochemical studies in relation to trace mineral status in camel. *M.V.Sc. thesis* submitted to the Rajasthan Agricultural University, Bikaner (Raj).

Fakhruddin (1987): Mineral and vitamin deficiencies in livestock of western Rajasthan. *Scientific programme and abstracts* [1(c) 13], National symposium on "Livestock production vis-a.-vis fodder and pasture development in western Rajasthan", 2-3 March 1987, Livestock Research Station, Rajasthan Agricultural University, Chandan (Jaisalmer-Raj).

Gahlot, A.K.(1997): Epizootiology and clinical aspects of surra in animals. *Short Course on Surra in animals* (March 17-21, 1997), Department of Parasitology, College of Veterinary and Animal Science, Rajasthan Agricultural University, Bikaner (Raj), pp30-35,

Kishore, R. (1998): Clinico-haemato-biochemical and therapeutic studies on bovine suffering from allotriophagia. *MV.Sc. thesis* submitted to the Rajasthan Agricultural University, Bikaner (Raj).

Mertia, R.S. (1987): Change in the botanical composition under different systems of grazing on rangelands in arid region of India. *Scientific programme and abstracts* [1 (a) 13], National symposium on "Livestock production vis-a-vis fodder and pasture development in western Rajasthan", 2-3 March 1987, Livestock Research Station, Rajasthan Agricultural University, Chandan (Jaisalmer-Raj).

Mittal, J.P. (1987): Impact of Indira Gandhi canal on livestock production in western Rajasthan. *Scientific programme and abstracts* [1(a)7], National symposium on "Livestock production vis-a-vis fodder and pasture development in western Rajasthan", 2-3 March, 1987, Livestock Research Station, Rajasthan Agricultural University, Chandan (Jaisalmer, Raj).

National Livestock Policy (1996): Report of the Steering group (NLP), Planning Commission, Govt, of India, New Delhi.

Pathak, K.M.L. (1997): Preface. *Short Course on Surra in animals* (March 17-21, 1997), Department of Parasitology, College of Veterinary and Animal Science, Rajasthan Agricultural University, Bikaner (Raj).

Radostits, A.M.; Blood, D.C. and Gay, C.C. (1994): Veterinary Medicine, Eighth Ed., Bailliere Tindall, London, U.K.

Sharma, S.N. and Gahlot, AK. (1997): *Veterinary Jurisprudence,* 4th ed., NBS Publishers, Bikaner (Raj).

Singh, A.P. and Gahlot, AK. (1999): Studies on Sarcoptes scabei infestation in dromedary. Abstracts, lJlh International Conference of the World association for the advancement of Veterinary Parasitology, 15-19 August, 1999, Denis Centre for Experimental Parasitology, The Royal Veterinary and Agricultural University, Copenhagen (Denmark), pp C.6.10.

Singh, A.P.; Netra, P.R; Vashistha, M.S. and Sharma, S.N. (1994): Zinc deficiency in cattle. *Indian Journal of Animal Sciences,* 64:35-40.

Singh, A.P.; Vyas, S.K; Tanwar, RK; Sharma, S.N. and Gahlot, A.K (1999): Studies on diagnosis and therapeutic aspects of sub clinical mastitis in cattle by using ampicillin and cloxacillin combination (AC-VET). *Intas Polivet,* 1:13-19.

Tanwar, RK.; Kumar, A.; Sharma, S.N.; Yadav, I.S. and Mathur, P.D. (1983): Clinical findings and therapy of polio encephalomalacia in goats. *Indian! Vet. Med.,* 3:112-114.

Tanwar, RK; Malik, KS. and Gahlot, A.K. (1994): Polioencephalomalacia induced with amprolium in buffalo calves: Clinic-pathological findings. *Journal of Veterinary Medicine,* 41:396-404.

27

Diagnostic approach for the livestock affected drought situation

B.R.J. Mathur

Ex-Deputy Director, Animal Husbandry Department, Government of Rajasthan, Jodhpur-342 001

The western part of the Rajasthan remains one of the most drought affected area. However, Animal husbandry remains the main source of livelihood. This area is also home ground for some of the best breeds of cattle, goat, sheep camel and Horse in this nation. In local sayings also it is said that there is only one bumper crop one in 4-5 years. The literary meaning of the word MARWAR also means मरने वालो की जगह -- the land of "Warriors"

So it becomes all the more important from animal husbandry point of view for timely diagnosis and confirmation of commonly prevailing diseases to conserve and sustain the live stock in drought period.

The diagnostic approaches for the livestock affected during draught situation can be mainly classified under five following headings —

- Deficiency diseases due to nutritional stress i.e., deficiency of vit A, Phosphorus, Copper, Magnesium, Cobalt, Calcium.
- Poisoning- Like nitrate, cyanide, aflatoxin, urea, Lantana.
- Indigestion - Alkalosis, Acidosis, constipation, diarrheas, Pica and botulism.
- Physical factors -- Heat strokes, dehydration, and epidemiologic indications cold and pneumonia etc.

- Loss of Immunity -- primarily due to above factor and extreme climatic conditions due to opportunistic nature of infection micro-organizes whether they may be Bacterial / Viral/protozoa/Fungal/clamy dial/Rikketrial/Ecto parasites / Endo parasites etc.

NUTRITIONAL DISEASES

Acetonemia in cattle, Pregnancy toxemia in sheep

Etiology: A multi factorial disorder of energy metabolism.

Negative energy to hypoglycemia and ketonemia (accumulation in blood of acetoacetate, beta-hydroxybutyrate and their decarboxylation product acetone and isopropanol).

The disease in cattle and sheep occur in different part of the pregnancy-lactation cycle.

Epidemiology: Primary ketosis occur in well conditioned cows with high lactation potential, principally in the first month of lactation with a higher prevalence in cows with higher lactation number.

Secondary ketosis occur where other disease reduce feed intake.

The disease in sheep is associated with a falling plane of nutrition. Principally in the last month of pregnancy in ewes bearing twins and triplets but can be induced by other stress at this time.

Signs : Cattle show wasting with decrease appetite

Fall in body condition as milk production, Neurological sign.

Sheep encephalopathy with blindness muscle tremor, convulsion, metabolic acidosis.

Clinical pathology : Hypoglycemia, ketonemia, ketourea or elevated ketones in milk.

Necropsy finding : Non specific, Twin lambs and fatty liver.

Diagnostic confirmation : Ketonimia, Ketonurea elevated beta hydroxy butyrate (BHBA) in aqueous humor of dead sheep.

Disease caused by deficiency of vitamin 'A'

Etiology: Dietary deficiency of vit-A or its precursors.

Epidemiology: Common in cattle grazing dry pastures for long periods, occur when diet of hand fed animals is not supplemented with vit A.

Signs cattle: Night blindness, loss of body weight, convulsions followed by recovery, permanent blindness with dilated pupils and optic disc edema.

Pig : Convulsion, hind leg paralysis, Congenital defects.

Clinical pathology: Low level plasma vit A

Lesions: Squamous metaplasia of inter lobular duct of parotid gland, compression of optic nerve track and spinal nerve root, degenerations of testes.

Diagnostic confirmation: Low level of plasma vit A and squamous metaplasia of interlobular duct of parotid glands.

Mineral Deficiency

1. Calcium (Ca)

Etiology: Primary dietary deficiency of calcium uncommon secondary calcium deficiency due to marginal calcium intake and high P intake.

Epidemiology: Sporadic, Not common if diet adequate.

Signs: Poor growth and dentition. Tetany may occur in lactating ewes. Inappetance, stiffness, fracture of long bones.

Clinical pathology: Serum calcium, Phosphorus and radiography.

Lesions: Osteoporosis, low ash content of bone.

Diagnostic confirmation: Histology of bone and bone ash analysis.

2. Phosphorus (P)

Etiology : Usually a primary deficiency in diet, may be conditioned by Vit D deficiency.

Epidemiology: Primary P deficiency occurs world wide. Soils and crops commonly deficient in phosphorus.

Primary deficiency may occur in lactating dairy cattle in early lactation.

Occurs under range conditions in beef cattle and sheep.

Sign: Young animals grows slowly, develop rickets. Adults develop ostomalacia, unthriftiness, weight loss, reduce feed consumption, reluctance to move, fractures, impaired fertility.

Clinical pathology: Serum P, Phosphorus-content of diet

Lesions: Rickets and osteomalacia, lack of mineralization of bones.

Diagnostic confirmation: Histology of bone and bone ash analysis.

3. Cobalt (Co)

Etiology: Dietary deficiency of Cobalt resulting in a deficiency of vit B12

Epidemiology : Occurs primarily in cattle and sheep unsupplemented with Cobalt. World wide where soils are deficient in Cobalt. Associates with ovine white liver disease.

Sign : Inappetance, gradual loss of body weight, pica, marked pallor of the mucus membrane, wool and milk production decreases, decreasing lambing percent.

Clinical pathology: Co. or vit B12 concentration of liver, Co. Concentration methylmalonic acid in plasma and urine. Formiminoglutamic acid in urine, anemia.

Lesion: Emacialtion, nemosiderosis of spleen.

Diagnostic confirmation: Vit B12 and cobalt of liver.

4. Copper (Cu)

Etiology : Primary copper deficiency due to inadequate levels in diet. Secondary copper deficiency due to conditioning factor such as excess Molibidnum and sulfur in diet.

Epidemiology : Primary in young pastures with ruminants in spring and summer. Primary deficiency occurs in sandy soils and heavily weathered areas. Feed water supplies may contains Mo. sulfate, and iron salts which interfere with Copper metabolism.

Sign: Unthriftiness, change in hair colour chronic diarrhea, chronic lameness, neonatal ataxia, anemia, enlarge hoof.

Clinical pathology: Low serum and hepatic copper, anemia.

Lesions: Anemia, emaciation, hemosiderosis, osteodystrophy, myocardiopathy, demyelination in enzootic ataxia.

Diagnostic confirmation: low serum and hepatic copper and response to treatment.

5. Magnesium (Mg)

Etiology: Hypomagnesaemia resulting from inadequate Mg.in diet.

Epidemiology: Most commonly calves 2-4 month of age, on whole milk or milk replacer diets and poor roughage, diarrhea and chewing bedding or other coarse fiber may exagerate the deficiency.

Sign: Hypersensitivity to all external stimuli, fine muscle tremors progressing to spasticity and violent convulsions. Rapid course.

Clinical pathology: Serum Mg. levels below 0.8 Mg./dl, (Normal 2-2.5 Mg./dl) bone Ca:Mg. ratio above 90:1 (Normal 55:1)

Necropsy finding : Calcification of the spleen, diaphragm and endothelium of the arota and endocardium.

Diagnostic confirmation: Blood Mg. and response to treatment. Bone Ca:Mg. ratio.

Disease causes by major phytotoxins

Cyanogenic glycoside poisoning (HCN poisoning)

Etiology: Ingestion of specific plant species containing cyanogenic glycoside

Epidimiology: Seasonal and other variations in glycoside content lead to periods of enhanced toxicity of pasture.

Sign: Acute poisoning soon after access to toxic material, sudden onset of dyspnea, tremor, recumbency, convulsions, death within a few minute to hour.

Chronic poisoning: Neonatal goiter or arthrogryposis, adult: incoordination, dysuria, frequent urination.

Clinical pathology: Assay of HCN in body tissuses and fluids and in feed.

Necropsy finding: Bitter almond smell upon opening of carcass.

Diagnostic confirmation: Positive assay of rumen content.

Nitrate and nitrite poisoning

Etiology: Nitrate As Na, K Nitrate fertilizer, food preservative. Nitrite in nitrate rich, mouldy plant material or as nitrate converted in rumen in nitrite.

Epidemiology : Outbreaks due to access to abnormal materials. Hay or any plant growing profusely after a long drought or high nitrate contents soil.

Sign : Sudden onset of severe dyspnea, brown mucosae, short course.

Clinical pathology : Very high blood level of methemoglobin (9g/100 ml blood-lethal level)

Necropsy finding : Brown blood in some.

Diagnostic confirmation: Positive assay for nitrite in feed, blood or aquous humor.

Field test : A simple test "**Diphenylamine Blue Test**" is available for detecting nitrate in feed/urine/blood/rumen content samples.

Urea poisoning

Etiology: M. common during drought by feeding urea treated fodder- excessively or accidentally.

Clinical symptoms: 1. Abdominal pain. 2. Muscle tremors and incoordination. 3. Blood oozes out from orifices. 4. Severe struggling and bellowing before death.

Sample in living: 1. Blood and serum.

Post-mortem leisons and Samples: 1. Hemorrhages over heart. 2. Bright red colour undotted blood. 3. Congestion of liver spleen, kidney.

Post-mortem samples: 1. ingesta- Intestine at both ends, pieces of liver, kidney, spleen.

Treatment - 2% acetic acid solution.

Disease due to inorganic poisoning

Fluoride: Source-1. Old wells in use only during drought. In endemic areas of Barmer, Jalore, Sikar, Dungarpur, and other arid districts showed the evidence of skeletal flurosis. The highest prevalence of 60% was observed at 8.0 mg/l fluoride

Clinical symptoms: As a result of discoloration, the teeth of animals became weak and rough, loss of appetite, stiffness of bones, lameness. The poor nutritional status of animals further aggravates the susceptibility.

Samples in living : 1. Drinking water sample.

2. Blood and serum.

Post-mortem leisons and Samples: Depending upon the amount of fluorine ingested, affected teeth and bones appears with characteristical "Mottling" which indicates hypoplazia, hypocalcification, staining and abrasion. The most characteristic lesions are osteoporosis, exos-tosis of long bones and mandibular bones besides increase in osteoid tissue.

1. Aluminum sulfate 30 gm orally
2. Ca salt I/v.

INDIGESTION

Alkalosis results from over-feeding or poor quality of roughages. Where as **acidosis** develops due to over feeding of grain concentrates. The best way to diagnose besides the history is to examine ruminal, pH, Blood pH. Blood /urine chemistry can easily diagnose it.

PHYSICAL FACTORS

Heat strokes, cold epidemiological indicators (due to desert like climate), dehydration. These affections can be diagnosed easily by history and epidemiological indicators. For cold exposure and influenza routine blood picture will prove to be sufficiently effective.

LOSS OF IMMUNITY

Nutritional deficiencies and extreme clim-atic condition during draught leads to such a stressful condition, that all the infections microorganisms become opportunistically aggressive and aggravate the diseases. The main culprit here is the drought, which plays the role of catalysts and predisposes the animal to various diseases (Ex. **F.M.D., P.P.R., Sheep pox, Calf scour, C.C.P.P., B.Q., H.S., E.T., Brucellosis, T.B., Jonin disease).**

Ecto parasites - As far as **Ecto parasites (Ticks and mits)** are concerned the skin scrapings can be taken for lab. examination and confirmation. Here important point which is to be kept in mind that on the surface of scalp blade/knife wax is to be applied and skin is sufficiently cut in depth so as to ensure the inclusion of parasites from their burroughs made deep in skin.

Endo parasites - As regards to **endo parasites** their can be best choice that is feacal examination. small amount of faces (say 5-

10 gm) may be taken for examination. But as far as samples from dead animal are concerned they may be taken during post-mortem. eg. intestinal, cysts different larval route and stages during their lung, liver etc.

28

Diseases management and control in ruminants in Arid region

S. Kachhawaha*, N.V. Patil, A.K. Gahlot, B.K. Mathur and A.C. Mathur

**Krishi Vigyan Kendra, Central Arid Zone Research Institute, Pali-Marwar (Raj.) - 306 401*

Infectious diseases

Many of the infectious diseases have been recorded from the arid zone. Conventional vaccination programme is in vogue in this zone by various service departments of the state- Current programme is plagued with two main: problems. Firstly, it does not cover whole of the population of the species covered' due to less doses available/produced by the BP laboratories of the states. Secondly, some of the important species of arid zone like goat and camel are not usually covered under this programme. Therefore, this programme needs to be extended in whole of the susceptible population (Table 1).

Table 1. Conventional vaccinations against the diseases requires to be extended in whole of the susceptible population

S.N.	Vaccine available against the disease	Presently covered species/animal	Requires to be extended to animal	Required periodicity
1	Haemorrhagic Septicemia	Cattle, Buffaloes	Camels	At least twice a year and in endemic areas thrice a year
2	FMD	Cattle, Buffaloes	-	At least twice a year
3	Anthrax	Cattle	Sheep, Buffalo, Camel	At least once a year in Anthrax Belt
4	Entrotoxaemia	Sheep	Goat	-
5	Sheep Pox	Sheep	Goat	-

Furthermore, certain new vaccines available in India for a long time are required to be included in the regular schedule. Despite of the fact that epidemiological studies and data indicate occurrence and prevalence of such diseases, regular use of vaccines against them is not in practice in this area leading to economic losses to the farmers. Inclusion of these vaccines in regular control programmes will certainly improve productive and reproductive efficacy of animals, besides saving their life. Table 2 shows some of these vaccines required for the domesticated animals of the arid zone.

Table 2. New vaccines required to be included in regular schedule

Vaccine available against the disease	Presently covered species/ animal	Requires to be extended to animals	Required periodicity
Tetanus	Not as a regular practice	Horses, Camels (Gahlot *et al.* 2000), calves, kids, lambs afflicted with injuries	At birth and then once a year for two years
Theileriosis	Not as a regular practice	New born calves	As per recommended schedules of the manufacture
Rabies	Not as a regular practice	All animals bitten by stray dogs, whether identified or not	Post-exposure schedule of tissue culture vaccine
PPR	Not as a regular practice	Goat & Sheep	Once in three year

Non-infectious diseases: Apart from prophylaxis with vaccines, drugs effective in prevention of certain diseases are also available. Such chemoprophylactics should be used as a routine preventive measure against the diseases and in the animals mentioned in Table 3.

Table 3. Chemoprophylaxis of certain diseases required in arid zone

Chemoprophylaxis required against the diseases	Animals	Drugs	Periodicity
Coccidiosis	Calves, Kids, Lambs	Sulphadimidine/ Sulphacotrimoxazole	1-3 Months of Age
Shipping Fever (Pasteurellosis)	Adult Cows, Horses, Buffaloes, Camels	Long acting Oxytetracycline + Sulphas	Just before and after long transportation

Trypanosomiasis (Kapoor, 1997)	Camels (Camel Research, 1996), Cattle, Buffaloes	Quinapyramine Sulphate + Chloride	February and June every year
Helminthiasis	All Animals	Borad Spectrum Anthelmintics/ Ectendoparasiticide	Just before and after Monsoon
Mange (Singh *et al.*, 1999)	Camels (Camel Research, 1996)	Ivermectins and pesticide spray	Once a Year
Pica (Kishore, 1998)	Camels, Cattle, Buffaloes	Mineral Mixture, Salt and Deworming	Regular Supplementation
Ticks	Cattle	Pesticide Spray and dusting	at monthly interval during summer season

Production diseases: Milk fever and ketosis in cattle, buffaloes and goats, post-parturient haemoglobinuria in buffaloes, downer's cow syndrome and mastitis are important production diseases of this area. Milk fever in this area occurs mainly due to calcium deficiency without concurrent deficiency of magnesium. Therefore, the treatment of milk fever should constitute calcium boroguconate only. Most likely period of occurrence is within one week post-parturient. However, subacute calcium deficiency affecting production can occur during early lactation and late pregnancy. Ketosis occurs between 3-7 weeks post-parturient period. Diabetic ketoacidosis has also been recorded in cattle, buffalo and goat. Zinc insulin is to be administered with constant monitoring of blood and urine sugar levels. Pica in cattle, buffaloes, camels and goats in this area is another problem (Mathur *et al.*, 2005). Botulism because of pica in cattle has been recorded from this area (Kishore, 1998). Protein energy, salt, phosphorus calcium, vitamin A, D and E deficiencies, helminthiasis (mainly round worms) had been identified as the major causes of pica. Successful management of pica is being practiced on these lines (Gahlot, 2005).

Parasitic diseases: Though a host of parasitic diseases has been identified, it would be out of scope to deal each one of them. Nevertheless, some of them are a population problem and their management must be done as part of routine husbandry practice. Arid climate is not very conducive for some of the parasitic infestations like flukes and cestodes. But with increased availability of water for irrigation and the animal husbandry practice being either migratory or of intensive type leads to hitherto unreported type of parasitism in animals. Preventive measures are usually the only practicable approach in such cases as many of the clinical conditions might be either uneconomical or rather difficult to treat. Table 4 provides information about a few of them.

Table 4. Epidemiologically significant parasitic diseases

S.N.	Parasitic disease	Animals affected	Management
1	Round worms (in general)	Almost all species of the area, particularly, young calves (cattle, buffaloes), sheep, goats, camels	At least biannual deworming with a broad-spectrum anthelmintic like albendazole, fenbendazole, tetramisole, levamisole, closantel, Ivermectins etc.
2	Gid (Cerebral coeneurosis)	Goats o	Gid requires surgery. Prevention. as above
3	Hydatidosis	Sheep, Goats, Camels, Buffaloes	Prevention: as above
4.	Maggot wounds	All species	Timely ASD of the wound. Use of maggoticides-locally, ivermectins parentrally
5	Mange	All species, particularly camels (Singh & Gahlot, 1999), cattle, buffaloes, sheep, goats	Local miticidal application using herbal/pyrethrins/ pesticides as dust/spray + ivermectins parentrally
6	*Oestrus ovis* infestation	Sheep, goats	Ivermectins provide some degree of success

Toxicity: Cyanide, sodium chloride, nitrate and nitrite poisoning have been reported to occur in this zone since long (Sharma and Gahlot, 1997). Newer toxicity includes those with pesticides/ farm chemicals, feed additives, drugs, environmental pollutants etc. (Radostits *et al.,* 1994).

Pesticide and farm chemicals: With frequent use of pesticides/farm chemicals in agriculture in this zone, particularly in the areas under irrigation, risk of poisoning of livestock has increased many folds. This is mainly because most of the domesticated animals in such areas are reared as a component of mixed farming system. Accidental access to insecticides/pesticides/ farm chemical is the main cause of poisoning in not only animals but also human beings. Chlorinated hydrocarbons like chlordane, toxaphene, heptachlor etc., organophosphorus compounds and carbamates like endosul-phan, monocrotophos, chlorpyriphos etc. and chemicals like urea are among the commonest poisonings. Seed dressings, rodenticides and herbicides are another class of miscellaneous chemicals exposing animals accidentally. Pesticides like melathion, asuntol, pyrethrins, amitraz etc. are also used in control of ectoparasites of animals. Application of these compounds in the concentrations ions and applications higher than the recommended ones often results into toxicity (Radostits *et al.,* 1994).

Management and control

Diseases can not be effectively controlled without timely and accurate information about their occurrence (incidence and prevalence) or without the capability to measure the effects of control measures.

Disease management can be of three basic types: prevention, control and eradication. Prevention involves precluding the occurrence of disease in animals or populations where it does not already occur. Control involves reducing the frequency of occurrence or the severity of existing disease- Eradication involves total extirpation of a disease from an area or population.

Management may be attempted by manipulating the agent (risk factor), the host population, the environment, human activities, or by combinations of these methods. A detailed knowledge of the epizootiology of the disease is required in choosing the most appropriate method.

Every disease management program must have a clear rationale, objective and plan of action, as well as a predetermined method for assessing its efficacy.

Management of the causative agent/factor or its vector

In order to manage disease by attacking its cause it is necessary to know the cause and to have an effective method of dealing with it. Infectious agents may be divided into two classes. Endogenous agents are present in normal hosts without causing obvious disease or are ubiquitous in nature and produce disease opportunistically. Exogenous agent are not present in normal hosts, produce clearly defined disease when introduced and do not persist in the environment. It is impossible to eliminate endogenous agents even on a local basis because of many potential sources.

Infectious agents may be attacked free in the environment eg. by disinfection, within carriers, vectors or animal reservoirs or within the host (by chemotherapy). A disadvantage of attacking the agent within the host is that substantial injury may have occurred prior to treatment.

Any program based on widespread and repeated use of chemotherapy or biocides to attack living agents or their vectors is likely to stimulate development of acquired resistance in the target organism. Such methods lose their effectiveness rapidly.

Precisely targeted treatment, based on detailed knowledge of the etiology of the disease, that uses minimum amounts of a variety of chemicals and that is coordinated with other forms of management, is likely to be more effective than mass treatment with a single agent, in the long term.

Disease management through manipulation of the host population

- Dispersion of animals may be used to manage disease if the disease is localized to a specific site and if there is little risk of spreading the causative agent to new areas.
- Fences and other artificial barriers to restrict animal distribution or movement are expensive, require continual monitoring and maintenance and are subject to failure.
- Selective removal of diseased individual (test and slaughter)
- Reduction of animal numbers in an area either through dispersal or depopulation.
- Total elimination or eradication of an animal species for disease control will probably be difficult to accomplish except through a very sustained, intensive program.
- Disease management through treatment and immunization.
- Infectious disease could be prevented by Immunization or treated satisfactorily in the individual.
- Treatment may be appropriate for endangered species or intensively managed species where therapeutic agents can be delivered directly to the individual or small group.
- Treatment should be used to reduce the risk of translocating disease agents.
- Immunization is more likely to be effective for the control of diseases caused by microparasites (Viruses, bacteria, protozoa) than for diseases caused by macroparasites (arthropods, helminths)

- An effective vaccine should produce no significant disease, stimulate long-lasting immunity, protect against a wide range of varieties of the agent and be incapable of reversion to virulence
- Oral immunization, using baits, appears to hold the greatest promise for use in wild animals. Because such vaccines are distributed in the environment rather than introduced directly into the target animal, they must be non-pathogeneic for non target species and retain their immunogenicity under adverse conditions.
- The proportion of a population that must be immunized for disease control is directly related to the rapidity with which the disease is transmitted and to the population density.

Disease Management through environmental modification

- Disease management through manipulation of the environment requires a more detailed understanding of the etiology of the disease than is required for other forms of management.
- Habitat manipulation may not provide as rapid results are move direct means of dealing with disease agents but the effects are likely to be more long lasting.
- Nidality is very important in using habitat modification for disease management. The investigator identifies features that restrict the disease to a nidus and the manager modifies these to reduce or prevent the occurrence of disease.
- Environmental manipulation may be used to influence the causative agent, the host population, the population of other species involved in the disease, and inanimate factors involved in disease occurrence.
- Disease control, through habitat modification often involves making either or type decisions, in which the benefit of reduced losses to diseases must be balanced against reduction in some other beneficial factor(s).
- The probable effect of habitat modifications on occurrence and distribution of disease should be part of the environmental assessment of any such action.

Disease management through influencing human activities

- The public must understand the reason for the management and how it will be done if they are to support it. Public education should be a part of any major programme.
- Management programs should include continual feed-back to involved, and to the public, to maintain enthusiasm for the project.
- This is particularly important in the latter stages of successful campaigns, when disease management may cease to be a priority and the disease may be allowed to re-emerge.
- Animals should not be translocated without an understanding of the potential disease agents, present at both the site of origin and the release site.
- Animals that are translocated should always be held in strict quarantine with regular monitoring after they have been moved but before they are released.

Emergency and integrated management programs

- Emergency measures are usually only required for occurrences of new or exotic diseases.
- Containment followed by eradication, is the method of choice for dealing with focal disease emergencies.
- The success of an emergency disease control program depends on early detection of the disease and rapid implementation of management measures. The latter requires extensive prior planning and testing of the system.
- Concentric quarantine zones are usually established about foci of disease to minimize animal movement into the area and prevent movement out of the area. This may involve radical depopulation and construction of physical barriers such as fences.
- Alternate or fall-back strategies, such as immunization, must be planned in the event that the disease cannot be contained.
- Management programs that rely on a single technique often fail because of changes in the disease, the host, or the environment.

- Once the objectives of a program have been defined, all possible solutions should be evaluated singly and in combination.
- Short term, specific measures often result in a rapid decrease in disease occurrence but are usually expensive, transient in their effect and may have little impact on the general health of the population.
- Long-term, non specific measures may reduce the occurrence of disease over time, as well as having other beneficial effects on the population.
- Combinations of short and long term measures, applied either simultaneously or consecutively are often more effective than single methods.
- Public education should be included in every integrated disease management program.
- Continuous monitoring of the program is necessary so that new methods can be introduced as current methods become ineffective.

Assessing the effectiveness of a disease management programme

- Every disease management programme should include techniques to measure its effectiveness
- Methods for accessing the efficacy of a programme must be planed before the programme is commenced.
- Assessment usually involves a comparison among treated and un-treated populations or areas. General rules for choosing appropriate controls and samples apply.
- The parameters used for measuring success must be directly relevant to the objectives of the programme.
- Different parameters may have to be used by evaluating effectiveness at various stages of a programme.
- Data collection and analysis should begin before any management is done so that ineffective methods can be detected early in the programme and corrected.
- Assessment should continue throughout the programme so that changes in the effectiveness of methods can be detected.

- Methods for analyzing the relative benefit: cost and cost effectiveness of procedures are available and should be applied.
- Mathematical models may be very useful in assessing the effective of techniques.

References

1. Camel Research (1996): Sharma, Y. K. and Fakhruddin, Directorate of Research (Veterinary and Animal Science). Rajasthan Agricultural University. Bikaner (Raj),
2. Dongre, R.A. (2000): Haemato-biochemical studies in relation to trace mineral status in camel. M.V.Sc. thesis submitted to the Rajasthan Agricultural University, Bikaner (Raj).
3. Fakhruddin (1987): Mineral and vitamin deficiencies in livestock of western Rajasthan. Scientific programme and abstracts [1(c) 13], National Symposium on "Livestock production vis-a-vis fodder and pasture development in western Rajasthan", 2-3 March 1987, Livestock Research Station, Rajasthan Agricultural University, Chandan Jaisalmer (Raj.).
4. Gahlot, A.K. (2005). Animal health management in drought prone area. Paper presented in the seminar on “Sustainable livestock production in arid region”, March 22-23,, 2005. College of Veterinary and Animal Sciences, Rajasthan Agriculture University, Bikaner (Raj.).
5. Gahlot, A.K., Tanwar, R.K., Singh, A.P., Tinna, N.K. and Sharma, S.N. (2000): Observations on tetanus in camels. The Blue Cross book, 14:34-35.
6. Kapoor, M. (1997): Chemotherapy and Chemoprophylaxis of surra in animals. Short Course on Surra in animals, March 17-21, 1997. Department of Parasitology, College of Veterinary and Animal Science, Rajasthan Agricultural University. Bikaner (Raj.). Pp: 46-50.
7. Mathur B.K, Patil N.V., Mathur A.C. and Patel A.K.2005.Impaction of Rumen of desert buck due to Polythene bags- a case study. Livestock International 10(5):17-20.
8. Mathur, A.C., Mathur, B.K., Kaushish, S.K. and Patel, A.K. (2001). Effect of package of improved nutritional management practices on true anoestrous cases in buffaloes in rural arid Rajasthan. Paper presented at Ann. Conv. and Natl. Symp on ‘Fertility management of farm animals under adverse agroclimatic conditions’ 6-8 Oct 2001, Jodhpur (India) Souvenir & Abstracts p 39.

9. Radostits, O.M., Gay, C.C., Blood, D.C. and Hincholiff, K.W. (2000). *Veterinary medicine*, 9th Ed.W.B. Saunders Co. Ltd., London.
10. Radostits. O.M., Blood, D.C. and Gay, C.C. (1994). *Veterinary Medicine*, Eighth Ed., Bailliere Tindall. London (U.K.)
11. Sansom, B.P. (1973). Clinical problem in preventive medicine. *Britts. Vet. J.*, 129: 207-220.
12. Sharma, S.N. and Gahlot, A.K. (1997). *Veterinary, Jurissprudence*, 4th ed., NBS Publishers, Bikaner (Raj).
13. Singh, A.P. and Gahlot, A.K. (1999). Studies on sarcoptes scabei infestation in dromedary. Abstracts, 17th International Conference of the World association for the advancement of Veterinary Parasitology, 15-19 August, 1999, Denis Centre for Experimental Parasitology, The Royal Veterinary and Agricultural University, Copenhagen (Denmark), pp: C. 6.10.
14. Singh, A.P. and Gahlot, A.K. (2004). *Practical manual of clinical veterinary medicine II*, Department of Clinical Veterinary Medicine, CVAS, Bikaner.
15. Singh, A.P., Netra, P.R., Vashistha, M.S. and Sharma, S.N. (1994). Zinc deficiency in cattlo. *Indian J. of Animal Science*. 64:35-40.
16. Singh, A.P., Vyas, S.K., Tanwar. R.K. Sharma, S.N. and Gahlot, A.K. (1999). Studies on diagnosis and therapeutic aspects of subclinical mastitis in cattle by using ampicillin and cloxacillin combination (AC-VET). *Intas Polivet* 1:13-19.
17. Tanwar, R.K., Fakhruddin and Chahar, A. (2008). *Management of disease in livestock. Seminar on "Sustainable Livestock production in arid region"*. March 22-23, 2005, held at College of Veterinary and Animal Sciences, Rajasthan Agriculture University, Bikaner (Raj.).

29

Animal Insurance -A financial safeguard to livestock owners during natural calamities

Dr. V.S. Shekhawat

Deputy Manager, The New India Assurance Co. Ltd., Udaipur (Rajasthan)

What is Insurance

The business of insurance is related to the protection of the economic value of the assets. Every asset has a value and on untimely loss through an accident or natural calamity; the owner suffers. Insurance is a mechanism that helps to reduce such adverse consequences.

Purpose and Need of Insurance : Assets are insured because they are likely to be destroyed or made non-functional through an accidental occurrence which may or may not happen. There has to be an uncertainty about the risk otherwise is not insurable. Insurance neither protects the asset nor prevents its loss due to the peril (e.g. accident, flood, earthquake etc.). Insurance compensates the losses which may not be full compensation depending upon type of policy. Only economic or financial losses can be compensated. Insurance covers tangible and intangible assets (e.g. voice of singer).

Contract of Insurance

When the insured pays the premium and the insurer accept the risk, the contract of insurance is concluded. The policy issued by the insurer is the evidence of contract.

The insured pays premium and insurer "promises to indemnify" as per the terms and condition of policy.

Who can Insure

If the person is likely to suffer financially when the property is lost or damaged, he has insurable interest and can insure that property. The insurable interest may be by way of ownership or as financer or as tenant etc. and it should be present both at the commencement of insurance and at the time of loss.

Indemnity: The principle of indemnity arises under common law and requires that an insurance contract should be a contract of indemnity only and nothing more. The object of the principle is to place the insured after a loss in the same financial position as far as possible, as he occupied immediately before loss. The effect of this principle is to prevent the insured from making a profit out of this loss or gaining any benefit or advantage.

The object of a contract of insurance is o protect the financial interest of the insured in the subject matter of insurance. If the insured could recover any amount in excess of his loss, he would make a profit, and if this is permitted, there may be a temptation for the insured himself to bring about the losses so as to make a profit. This would bc against public policy.

The measure of indemnity for loss of or damage to the insured property is generally the intrinsic market value of the property at the place and on the date of loss or damage.

Rural Insurance: The rural population of India is dependent on agriculture and animal husbandry. The losses to crops or livestock are beyond the capacity of rural poor to bear.

Obligation of Insurers to the Rural Sector : Keeping in view this fact Section 32 B and 32 C of the insurance Act 1938 states that insurers are obliged to underwrite 2%, 3% and 5% of total gross premium in the first, second and third year onward, the business in the rural sector including insurance for crops.

The rural sector has been defined as any place which as per the last census, has a population not more than 5000, density of population not more than 400 per square KM and at least 75% of the male working population engaged in agriculture.

Rural Policies : Rural policies comprise insurance of—

Livestock - e.g. cattle, sheep, goat, camel etc.

Various food crops - cereals, millets, pulses, oilseeds, cotton seeds, sugarcane, Potato, etc. This done by the Agriculture Insurance Company.

Plantation and Horticultural crops – e.g. rubber, grapes, etc. Floriculture

Property - e.g. agriculture pumpsets, tractors, etc.

Persons - e.g. Janta and Gramin personal Accident

Sub-animals–e.g. silkworm, honeybee, brackish water prawn, etc.

The Role of Government

The Govt. of India has launched various programmes for the benefit of small farmers, marginal farmers, agricultural labourers, etc. Since 1980 all these schemes were integrated into Integrated Rural Development Programme (IRDP) funded by Central and State Govts.

Special insurance schemes are framed to protect the beneficiaries of IRDP Projects. Under these policies, the rates of premium are lower and claim procedure is simplified. Whenever the word 'scheme' is used hereafter, it refers to these special policies.

CATTLE INSURANCE

Highlights

- This scheme covers the following whether indigenous, exotic or cross-bred.
- Milch Cows and Buffaloes
- Calves / Heifers
- Stud Bulls
- Bullocks (Castrated Bulls) and Castrated Male Buffaloes,
- Animals within a specified age group are accepted under the Standard Insurance Scheme.
- Sum Insured under the policy will be the Market Value of the animal.
- Indemnity under the policy will be the sum insured or market value prior to illness whichever is less. The indemnity is limited to 75% of Sum Insured in case of a PTD claim.
- The basic premium rate per annum is 4% of the Sum Insured. Long term policies are also issued with long term discounts.

- The premium rates under the policy are concessional for covering animals under government subsidized schemes.
- Group Discounts are also available.

Insurance Coverage

- The policy shall give indemnity for death due to.
- Accident (Inclusive of fire, lightning, flood, inundation, storm, hurricane, earthquake, cyclone, tornado, tempest and famine).
- Diseases contracted or occurring during the period of this policy.
- Surgical Operations.
- Riot and Strike.
- The Policy can also be extended to cover PTD on payment of extra premium;
- Permanent Total Disability which, in the case of Milch Cattle result in permanent and total incapacity to conceive or yield milk.
- PTD which in the case of Stud Bulls results in permanent and total incapacity for breeding purpose.
- In case of Bullocks, Calves / Heifers and Castrated male buffaloes results in permanent and total incapacity for the purpose of use mentioned in the proposal form.

Major Exclusions

(A) Common Exclusions

Malicious or willful injury or neglect, overloading, unskillful treatment or use of animal for purpose other than stated in the policy without the consent of the Company in writing.

Accidents occurring and /or Disease contracted prior to commencement of risk.

Intentional slaughter of the animal except in cases where destruction is necessary to terminate incurable suffering on humane consideration on the basis of certificate issued by qualified Veterinarian or in cases where destruction is resorted to by the order of lawfully constituted authority.

Theft and clandestine sale of the insured animal.

War, invasion, act of foreign enemy, hostilities (whether war be declared or not), civil war, rebellion, revolution, insurrection, mutiny, tumult, military or usurped power or any consequences thereof or attempt threat.

Any accident, loss, destruction, damage or legal liability directly or indirectly caused by or contributed to by or arising from nuclear weapons.

Consequential loss of whatsoever nature.

Transport by air and sea.

Any non-scheme claim arising due to diseases contracted within 15 days from the date of risk are not covered.

(B) Specific Exclusions

All the claims received without ear tag.

Documents to Effect Insurance Coverage

Proposal Form

Veterinary Health Certificate from a qualified Veterinarian giving the age, identification marks, health, and market value of the animal in the prescribed format.

Identification of Animal

- All insured animals should be suitably identified by natural Identification marks and color should be clearly noted in the proposal form and Veterinarian's Report.
- Ear tags made of suitable material are applied to the ear of the animals and the code number is entered into the Veterinary Health Certificate.
- Photographs of animals may be insisted in case of high value animal.

Claim Procedure

- In the event of death of an animal, immediate intimation should be sent to the Insurers and the following requirements should be furnished:
- Duly completed claim form.
- Death Certificate obtained from qualified Veterinarian on Company's form.

- Postmortem examination report if required by the Company.
- Ear Tag applied to the animal should be surrendered. The condition of' No Tag- No claim' will be applied if the tag is not surrendered.
- Claim Procedure For PTD Claim
- A certificate from the qualified Veterinarian to be obtained.
- The animal will be inspected by the company's Veterinary Officer also.
- . Complete chart of treatment, medicines used, receipts, etc., should be submitted.
- Admissibility of claim will be considered after two months of Veterinary Doctor / Company Doctor's report.
- The indemnity is limited to 75% of Sum Insured.

Sheep and Goat Insurance

Highlights

All indigenous, crossbred and exotic Sheep and Goat will be covered under this Scheme.

Scope

The policy shall provide indemnity against death of sheep and goats due to accident Including Fire, Lightning, Flood, Cyclone, Famine, Earthquake, landslide, Strike, Riot or diseases contracted or occurring during the period of insurance.

Exclusion

Common Exclusion

Same as those for cattle insurance.

Specific Exclusion

Enterotoxaemia, Sheep Pox, Goat Pox, Rinderpest, FMD, Anthrax, H.S., B.Q. These diseases are covered by the policy if the animal is successfully inoculated (protected) and necessary Veterinary Certificates are supplied to the Company.

Sum Insured

The market value of sheep and goats varies from breed to breed, from area to area and from time to time. The examining

Veterinarians recommendations shall be considered as the proper guide for acceptance of insurance as well as for settlement of claims.

Sum insured will not exceed 100% of market value.

Claim Procedure

In the event of death immediate intimation should be given to the Company and the Insured should furnish the following documents and required information.

Duly completed claim form.

Death certificate from a Veterinarian on Company's form.

Post mortem examination report if required by the Company.

Ear tag wherever applicable.

OTHER POLICIES WHICH ARE BENEFICIAL TO THE RURAL MASSES

Cattle Insurance

Sheep and Goat Insurance

Horse / Pony / Mule / Donkey Insurance

Camel Insurance

Poultry Insurance Scheme

Agricultural Pumpsets Insurance Scheme

Animal Driven Cart Insurance

Hut Insurance

Gobar Gas Insurance Scheme

Lift Irrigation Insurance

Janata Personal Accident

Gramin Personal Accident Policy

Composite Package For Tribals

Farmers Package Insurance Scheme

Insurance: provides safeguards and financial support in case of livestock death or disability through an accident or illness or natural calamities. The purpose of insurance is to safeguard against such misfortunes. Thus insurance is a economic and social security tool.

Author Index